雅力教育集团
RAY EDUCATION GROUP

◆ 雅力教育集团介绍

　　雅力教育集团，作为中国国际教育的前锋，是致力于国际学校投资运营、国际课程研究开发、素质教育项目开发于一体的全球教育集团新锐．作为中国首批战略投资海外教育资产的机构，集团拥有两所英国百年贵族私立学校（英国阿德科特学校和英国米德尔顿公学）．集团在中国的上海、苏州、广州、合肥、武汉等十余个城市开设K12国际学校、国际课程中心及幼儿园，并在英、美、澳、加等国设有海外服务中心，至今已输送数百名学生进入牛津、剑桥、耶鲁、普林斯顿等世界前50名校．

　　集团基于"扎根中国 布局全球"的发展思维，以雅力卓越学院为基础平台，以数字技术为重要支撑，整合国内外校区，链接全球优质教育资源，打造雅力全球校园，培养具有传承、创新、独立、领袖精神，能够行走世界的中国人！

◆ Introduction of Ray Education Group

Ray Education Group is considered one of the pioneers of international education in China. It is dedicated to the investment and operation of K12 international schools, as well as international curriculum research and whole-person education programme development. Ray Education Group is the first private education institution in China that has made a strategic investment and purchase of two century-old British Elite Independent schools (Adcote School and Myddelton College) in the United Kingdom. Ray Education Group operates international schools, international curriculum centers and kindergartens in more than 10 cities in China, including Shanghai, Suzhou, Guangzhou, Hefei, Wuhan, to name a few. It also runs oversea service centers in the United Kingdom, the United States, Australia, Canada and other countries. So far, Ray Education Group has successfully sent hundreds of students to the world's top 50 universities, including the University of Oxford, the University of Cambridge, Yale University, and Princeton University.

Ray Education Group, assisted with Ray Excellent Academy and digital technology, has a development and strategic vision of "Chinese Root, Global Reach". By integrating quality teaching products and resources among Ray schools and globally, Ray Education Group is building a global campus to cultivate Chinese youth who can walk freely in the world with the merit of inheritance, innovation, independence and leadership.

雅力教育集团

雅力卓越学院

剑桥A-level国际课程备考丛书
Cambridge A-level International Course Preparation Book Series

剑桥A-level
纯数1课程精解

Cambridge A-level Pure Mathematics 1 Upgrading
Your Key to G5 and Oxbridge

冲刺G5、牛剑必备宝典

本册主编： 周则鸣
本册翻译： 严朝红
本册校对： 严朝红　吕倩倩　周高全　潘　丽
　　　　　　孙海亮　查　蔓　童广悦

雅力教育集团
RAY EDUCATION GROUP

主编简介

周则鸣（Steve Che-Ming, Chou）在温哥华就读高中时学习 IB 课程．本科及研究生毕业于加拿大不列颠哥伦比亚（UBC）大学，获得电气工程、Honors 数学及金融数学学位．曾任职于 UBC 大学数学系，2009 年来到中国从事国际教育．任教期间还担任多年的加拿大和美国（AMC）奥林匹克数学培训指导老师，为世界各大名校输送了大批优秀的学生．2018 年 10 月加入上海阿德科特学校．在上海阿德科特学校的第一届 10 位毕业生中，有 4 位获得牛津、剑桥面试邀请并且全部顺利通过面试录取．因此，他被业内和学生们誉为"牛剑爸爸"．

作为一位读过、考过、教过国际课程并具有多年研究指导工作经验的专业教师，他对各类国际课程体系及课程设置等相关信息有全面的了解和独到的见解，并且勇于开发创新，能够让中国学生有效、迅速地接受国际课程，克服自身英语学习的瓶颈．

内容提要

A-level（General Certificate of Education Advanced Level）是英国普通高级中学教育水平认证，被称为世界范围内的"黄金教育体系"．A-level 国际高考文凭是目前世界上广泛认可的高中文凭之一，特别是在申请牛津、剑桥等世界顶级名校时，A-level 的重要性无可比拟．

《剑桥 A-level 纯数 1 课程精解》是剑桥 A-level 国际课程备考丛书中数学第一册，具体包括二次方程式、函数、几何坐标、弧度法、三角函数、二项式定理、数列和级数、微分、微分的应用、反导数和积分等内容．本书依循剑桥数学考试的官方考点和扩展教科书的核心知识点，为学生提供额外的习题训练和实践活动．本书为每个知识点提供独到生动的解读，深入浅出地阐释背后的数学原理，并为每道习题提供正确、详尽的解答，适合作为教学辅助材料、家庭作业习题册或学生自学教材．

序言
Foreword

20 世纪 90 年代末,剑桥大学 A-level 课程进入中国大陆,迄今已经近 20 年。A-level 课程被称为最适合中国学生的国际课程,它包含 70 多门科目供学生选择。在众多科目中,数学科目在中国可以说是最热门的学科,这不仅因为中国学生在数学方面的优势,还因为数学科目最后可以以基础数学和进阶数学两门来申请大学,这也为很多中国学生冲刺牛津、剑桥、英国 G5 大学带来优势.

虽然 A-level 课程进入中国近 20 年,但是,因为它是由剑桥大学统一出题、统一阅卷,剑桥大学考试局统一制定教学大纲和考试大纲,每所学校使用的教材也都是原版引进教材,在中国市场少有关于 A-level 课程的精解教材或是练习册,每所学校的教学复习和模拟测试基本上都是依靠剑桥的历年真题.因此,出版一套符合中国学生学习需求的 A-level 精解教材在行业内有很长时间的呼声.这需要有大量的实践积累,需要有经验的教学团队,不断从教学实践中总结经验,这无疑需要一个漫长的过程.

雅力教育集团从 2006 年成立至今,一直致力于海外留学移民业务、国际合作办学和国际课程研发,在国际教育领域深耕十余年.此书由雅力教育集团旗下的国际学校教师编著,领衔的周则鸣老师是我多年的同事,他从事 A-level 教学十多年,培养出很多数学专业的优秀学子进入牛津、剑桥、英国 G5 等世界知名大学.凭借作者团队多年的教学积累,我相信这本书会使很多学习 A-level 课程的中国学生获益.

上海阿德科特学校校长　车艳丽

2020 年 2 月

目录 Content

1　Quadratics　第 1 章　二次方程式 **001**

1.1　Completing the Square　配方法 .... / 002
1.2　Quadratic Formula　二次方程求根公式 .... / 003
1.3　Graph of Quadratic Function　二次函数方程式图像 .... / 005
1.4　Solving Quadratic Inequality　解二次不等式 .... / 008
1.5　Intersection between Line and Parabola　直线与抛物线相交 .... / 009
1.6　Extended Material　拓展内容 .... / 013
1.7　Recursive Formula for the Sum of nth Power of Roots　根的 n 次幂之和的递归公式 ...... / 015

2　Functions　第 2 章　函数 **016**

2.1　The Vertical Line Test　垂直线测试 .... / 018
2.2　Symmetry of Functions　函数的对称性 .... / 019
2.3　One-to-one Function　一一对应函数 .... / 020
2.4　The Horizontal Line Test　水平线测试 .... / 021
2.5　Composite Function　复合函数 .... / 021
2.6　Inverse Function　反函数 .... / 024
2.7　Transformation of Function　函数变形 .... / 027
2.8　Stretch/Reflection　拉伸/翻折 .... / 029
2.9　Special Reflection in Straight Lines　直线上的特殊翻折 .... / 030
2.10　Combining Transformation　组合变形 .... / 030

3　Coordinate Geometry　第 3 章　几何坐标 **040**

3.1　Line and the Gradient　直线与斜率 .... / 040
3.2　Parallel and Perpendicular Lines　平行线与垂线 .... / 042
3.3　Length of Line Segment and Midpoint　线段的长度与中点 .... / 044
3.4　The Properties of Circle　圆的性质 .... / 046

3.5　The Cartesian Equation of Circle　圆的笛卡尔方程 ……………………………… / 051
3.6　Standard Equation of a Circle　圆的标准方程 …………………………………… / 051
3.7　Intersection between Line and Circle　直线与圆相交 ……………………………… / 053
*3.8　Intersection between Two Circles　圆与圆相交 ………………………………… / 054

4　Circular Measure　第 4 章　弧度法 ……………………………………………… 065

4.1　Arc Length　弧长 …………………………………………………………………… / 066
4.2　Area of Sector　扇形面积 ………………………………………………………… / 067
4.3　Further Problems Involving Arcs and Sectors　弧与扇形的拓展题目 …………… / 068

5　Trigonometry　第 5 章　三角函数 ………………………………………………… 080

5.1　Angle Measurement　角的测量 …………………………………………………… / 081
5.2　Cosine is an Even Function and Sine is an Odd Function　余弦是偶函数，正弦是奇函数
　　　……………………………………………………………………………………… / 082
5.3　Complementary Angle Identities　余角等式 ……………………………………… / 083
5.4　Supplementary Angle Identities　补角等式 ……………………………………… / 083
5.5　Trigonometric Identities　三角恒等式 …………………………………………… / 087
5.6　Graph of Trigonometric Functions　三角函数的图像 …………………………… / 089
5.7　The Sine Curve　正弦曲线 ………………………………………………………… / 090
5.8　The Cosine Curve　余弦曲线 ……………………………………………………… / 092
5.9　The Tangent Curve　正切曲线 ……………………………………………………… / 094
5.10　Inverse Trigonometric Functions　反三角函数 ………………………………… / 097
5.11　The Inverse Sine (or Arcsine) Function　反正弦函数 …………………………… / 097
5.12　The Inverse Tangent (or Arctangent) Function　反正切函数 …………………… / 099
5.13　The Inverse Cosine (or Arccosine) Function　反余弦函数 ……………………… / 101
5.14　Solving Trigonometric Equations　解三角方程 ………………………………… / 103

6　Binomial Theorem　第 6 章　二项式定理 ………………………………………… 110

6.1　Pascal's Triangle　帕斯卡三角形 ………………………………………………… / 110
6.2　Pascal's Formula　帕斯卡公式 …………………………………………………… / 113
*6.3　Multinomial Expansion　多项式展开 …………………………………………… / 119

7　Sequence and Series　第 7 章　数列和级数 ………………………………………… 123

7.1　Arithmetic Sequence　等差数列 …………………………………………………… / 123
7.2　Arithmetic Series　等差级数 ……………………………………………………… / 126
7.3　Geometric Sequence　等比数列 …………………………………………………… / 128

7.4　Geometric Series　等比级数 …… / 130

7.5　Infinite Geometric Series　无穷等比级数 …… / 132

8　Differentiation　第 8 章　微分 …… 136

*8.1　Limit　极限 …… / 136

8.2　Limit at Infinity for Rational Function　有理函数在无穷大的极限 …… / 138

8.3　Tangent Lines and Gradients　正切线及斜率 …… / 138

8.4　The Derivative　导数 …… / 139

8.5　Other Notations for Derivative　导数的其他符号 …… / 140

8.6　Basic Differentiation Rules　基本微分法 …… / 141

8.7　The Chain Rule and Higher Order Derivative　链式法则与高阶导数 …… / 145

9　Applications of Differentiation　第 9 章　微分的应用 …… 151

9.1　Increasing and Decreasing Function　递增与递减函数 …… / 151

9.2　Maximum/Minimum and Stationary Points　最大值/最小值及驻点 …… / 154

9.3　Related Rate　相关变化率 …… / 159

9.4　Optimization Problems　优化问题 …… / 161

10　Anti-derivative and Integration　第 10 章　反导数和积分 …… 166

10.1　Indefinite Integral　不定积分 …… / 167

*10.2　Riemann Sum　黎曼积分 …… / 172

10.3　Definite Integral　定积分 …… / 174

10.4　Area between Curves　曲线之间的面积 …… / 178

10.5　Improper Integrals　广义积分 …… / 181

10.6　Volume of Revolution　旋转体体积 …… / 185

校长推荐 …… 192

学生推荐 …… 193

Quadratics

第 1 章　二次方程式

Definition 定义

A quadratic equation is an equation written in the form of $ax^2 + bx + c$, where a, b and c are constants and $a \neq 0$.

二次方程式的标准表达式为 $ax^2 + bx + c$,其中,a, b, c 为常数,且 $a \neq 0$.

We can solve the quadratic equation by factorization or completing the squares.
We shall know the simple expansions and factorizations listed below:
$(x + y)^2 = x^2 + 2xy + y^2$, $(x - y)^2 = x^2 - 2xy + y^2$, $x^2 - y^2 = (x + y)(x - y)$.
Now, let's start with a few examples to show how the factorization process works.

Example 例 1.1

Solve the following equations.

(a) $x^2 + 5x + 4 = 0$; (b) $5x^2 + 19x + 12 = 0$; (c) $\dfrac{3x^2 + x - 10}{x^2 - 7x + 10} = 0$; (d) $\dfrac{x^2 - 9}{x + 3} = 0$.

Solution 解

(a) $x^2 + 5x + 4 = 0 \Rightarrow (x + 1)(x + 4) = 0 \Rightarrow x = -1 \text{ or } -4$.

(b) $5x^2 + 19x + 12 = 0 \Rightarrow (x + 3)(5x + 4) = 0 \Rightarrow x = -3 \text{ or } \dfrac{-4}{5}$.

(c) $\dfrac{3x^2 + x - 10}{x^2 - 7x + 10} = 0 \Rightarrow \dfrac{(x + 2)(3x - 5)}{(x - 2)(x - 5)} = 0$, since there are no common factors; hence the numerator must be equal to 0. So $x = -2 \text{ or } \dfrac{5}{3}$.

(d) $\dfrac{x^2-9}{x+3}=0 \Rightarrow \dfrac{(x+3)(x-3)}{x+3}=0$, since there is a common factor $(x+3)$; hence we have to cancel out the factor. Therefore, we would have $(x-3)=0$. So $x=3$ only. ∎

Definition 定义

The root of a quadratic equation ax^2+bx+c is the value $x=\alpha$ such that $a\alpha^2+b\alpha+c=0$.

二次方程式 ax^2+bx+c 的根为 $x=\alpha$，则 $a\alpha^2+b\alpha+c=0$。

For any quadratic equation, we always have two roots. It can be two distinct real roots, repeated root or no real roots (complex roots).

任何二次方程式总是有两个根。它可以是两个不同的根、两个相同的根，也可以没有实根（即复数根）。

1.1 Completing the Square 配方法

Sometimes a quadratic equation is not easy to carry out the factorization. In such cases, we may consider the use of completing the square. Since we know $(x \pm y)^2 = x^2 \pm 2xy + y^2$, so the steps for completing the square are as the following:

Suppose we are given an equation $ax^2+bx+c \Rightarrow a\left(x^2+\dfrac{b}{a}x\right)+c \Rightarrow a\left(x^2+\dfrac{b}{a}x+\dfrac{b^2}{4a^2}-\dfrac{b^2}{4a^2}\right)+c$

$\Rightarrow a\left(x^2+\dfrac{b}{a}x+\dfrac{b^2}{4a^2}\right)-\dfrac{b^2}{4a}+c \Rightarrow a\left(x+\dfrac{b}{2a}\right)^2-\dfrac{b^2}{4a}+c.$

Example 例 1.2

Express the following in the form of $p(x-q)^2+r$, where p, q and r are constants.

(a) $2x^2+8x+1$; (b) $5x^2+10x+3$; (c) $3x^2+x-10$; (d) $-7x^2+3x-1$.

Solution 解

(a) $2x^2+8x+1 = 2(x^2+4x)+1 = 2(x^2+4x+4-4)+1 = 2(x^2+4x+4)-8+1 = 2(x+2)^2-7.$

(b) $5x^2+10x+3 = 5(x^2+2x)+3 = 5(x^2+2x+1-1)+3 = 5(x^2+2x+1)-5+3 = 5(x+1)^2-2.$

(c) $3x^2+x-10 = 3\left(x^2+\dfrac{1}{3}x\right)-10 = 3\left(x^2+\dfrac{1}{3}x+\dfrac{1}{36}-\dfrac{1}{36}\right)-10$

$= 3\left(x^2+\dfrac{1}{3}x+\dfrac{1}{36}\right)-\dfrac{1}{12}-10 = 3\left(x+\dfrac{1}{6}\right)^2-10\dfrac{1}{12}.$

(d) $-7x^2+3x-1 = -7\left(x^2-\dfrac{3}{7}x\right)-1 = 7-\left(x^2-\dfrac{3}{7}x+\dfrac{9}{196}-\dfrac{9}{196}\right)-1$

$= -7\left(x^2-\dfrac{3}{7}x+\dfrac{9}{196}\right)+\dfrac{9}{28}-1 = -7\left(x-\dfrac{3}{14}\right)^2-\dfrac{19}{28}.$ ∎

Example 例 1.3

Use completing the square to solve the equation $\dfrac{5}{x+2} + \dfrac{2}{x+3} = 3$. Leave your answers in surd form.

Solution 解

$\dfrac{5}{x+2} + \dfrac{2}{x+3} = 3 \Rightarrow \dfrac{5(x+3) + 2(x+2)}{(x+2)(x+3)} = 3 \Rightarrow 5(x+3) + 2(x+2) = 3(x+2)(x+3)$

$\Rightarrow 5x + 15 + 2x + 4 = 3(x^2 + 5x + 6) \Rightarrow 3x^2 + 15x + 18 - 7x - 19 = 0$

$\Rightarrow 3x^2 + 8x - 1 = 0 \Rightarrow 3\left(x^2 + \dfrac{8}{3}x\right) - 1 = 0$

$\Rightarrow 3\left(x^2 + \dfrac{8}{3}x + \dfrac{16}{9} - \dfrac{16}{9}\right) - 1 = 0 \Rightarrow 3\left(x^2 + \dfrac{8}{3}x + \dfrac{16}{9}\right) - \dfrac{19}{3} = 0$

$\Rightarrow 3\left(x + \dfrac{4}{3}\right)^2 = \dfrac{19}{3} \Rightarrow \left(x + \dfrac{4}{3}\right)^2 = \dfrac{19}{9} \Rightarrow x = \dfrac{-4}{3} \pm \dfrac{\sqrt{19}}{3}$.

Example 例 1.4

Find the real solutions of the following equations.

(a) $(x^2 - 7x + 11)^4 = 1$; (b) $(3x^2 + 5x - 7)^4 = 1$.

Solution 解

(a) $(x^2 - 7x + 11)^4 = 1 \Rightarrow (x^2 - 7x + 11)^2 = \pm 1$. Since we only want real solution, so
$(x^2 - 7x + 11)^2 = 1 \Rightarrow x^2 - 7x + 11 = \pm 1 \Rightarrow x^2 - 7x + 12 = 0$ or $x^2 - 7x + 10 = 0$
$\Rightarrow (x-3)(x-4) = 0$ or $(x-2)(x-5) = 0 \Rightarrow x = 3, 4$ or $2, 5$.

(b) $(3x^2 + 5x - 7)^4 = 1 \Rightarrow (3x^2 + 5x - 7)^2 = \pm 1$. Since we only want real solution, so
$(3x^2 + 5x - 7)^2 = 1 \Rightarrow 3x^2 + 5x - 7 = \pm 1 \Rightarrow 3x^2 + 5x - 8 = 0$ or $3x^2 + 5x - 6 = 0$
$\Rightarrow (x-1)(3x+8) = 0$ or $3\left(x^2 + \dfrac{5}{3}x + \dfrac{25}{36} - \dfrac{25}{36}\right) - 6 = 3\left(x + \dfrac{5}{6}\right)^2 = \dfrac{97}{12}$
$\Rightarrow x = 1, \dfrac{-8}{3}$ or $\dfrac{-5 \pm \sqrt{97}}{6}$.

1.2 Quadratic Formula 二次方程求根公式

We can solve a quadratic equation by using a more convenient way which is known as the quadratic formula. The quadratic formula is derived from completing the square.

我们可以通过使用一个更简单的办法来解二次方程,这个方法称为二次方程式求根公式,它从配方法演化而来。

Consider an equation $ax^2 + bx + c = 0$ where $a \neq 0$, then

$$a\left(x^2 + \frac{b}{a}x\right) + c = 0 \Rightarrow a\left(x^2 + \frac{b}{a}x + \frac{b^2}{4a^2} - \frac{b^2}{4a^2}\right) + c = 0 \Rightarrow a\left(x^2 + \frac{b}{a}x + \frac{b^2}{4a^2}\right) - \frac{b^2}{4a} + c = 0$$

$$\Rightarrow a\left(x + \frac{b}{2a}\right)^2 = \frac{b^2 - 4ac}{4a} \Rightarrow \left(x + \frac{b}{2a}\right)^2 = \frac{b^2 - 4ac}{4a^2} \Rightarrow x = \frac{-b \pm \sqrt{b^2 - 4ac}}{2a}.$$

Theorem 1.1　定理 1.1

A quadratic equation $ax^2 + bx + c = 0$ where $a \neq 0$, then the roots of the equation are $x = \dfrac{-b \pm \sqrt{b^2 - 4ac}}{2a}$.

有二次方程 $ax^2 + bx + c = 0$，其中 $a \neq 0$，则该方程的根为 $x = \dfrac{-b \pm \sqrt{b^2 - 4ac}}{2a}$。

Proof　证明

It is proved above.

The expression $b^2 - 4ac$ in the quadratic formula is known as the discriminant, we denoted it by $D = b^2 - 4ac$. The discriminant can help us to determine the roots of the quadratic equation. When

$D > 0 \Rightarrow$ two distinct real roots;

$D = 0 \Rightarrow$ repeated root;

$D < 0 \Rightarrow$ no real roots.

Example 例 1.5

Solve the following equations.

(a) $3x^2 - 5x + 1 = 0$; (b) $9x^2 - 6x + 1 = 0$; (c) $x^2 + x + 1 = 0$.

Solution 解

(a) $x = \dfrac{5 \pm \sqrt{25 - 12}}{6} = \dfrac{5 \pm \sqrt{13}}{6} \Rightarrow$ Two distinct real roots.

(b) $x = \dfrac{6 \pm \sqrt{36 - 36}}{18} = \dfrac{1}{3} \Rightarrow$ One repeated root.

(c) $x = \dfrac{-1 \pm \sqrt{1 - 4}}{2} = \dfrac{-1 \pm \sqrt{-3}}{2} \Rightarrow$ No real solutions.

1.3 Graph of Quadratic Function 二次函数方程式图像

The graph of a quadratic function $f(x) = ax^2 + bx + c$ is always a parabola when $a \neq 0$. With $a > 0$, the parabola opens upwards. With $a < 0$, the parabola opens downwards. By use of completing the square, we know $f(x) = ax^2 + bx + c = a\left(x + \dfrac{b}{2a}\right)^2 + \left(c - \dfrac{b^2}{4a}\right) = a(x - h)^2 + k$; therefore, the vertex of the parabola occurs at (h, k).

二次函数方程式 $f(x) = ax^2 + bx + c$ 的图像永远是一条抛物线 $(a \neq 0)$. 当 $a > 0$ 时,抛物线开口向上;当 $a < 0$ 时,抛物线开口向下. 通过配方法,可知 $f(x) = ax^2 + bx + c = a\left(x + \dfrac{b}{2a}\right)^2 + \left(c - \dfrac{b^2}{4a}\right) = a(x - h)^2 + k$. 所以,这条抛物线的顶点是 (h, k).

The following figures show the variety of different cases (Fig. 1.1).

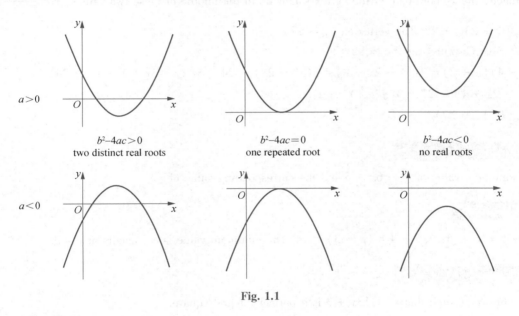

Fig. 1.1

Note 注意

When $a > 0$, the parabola has a minimum value which occurs at the vertex.

When $a < 0$, the parabola has a maximum value which occurs at the vertex.

The graph of the parabola is symmetry about the vertical line $x = h = \dfrac{-b}{2a}$.

当 $a > 0$ 时,抛物线的顶点是最小值;

当 $a < 0$ 时,抛物线的顶点是最大值;

抛物线是关于 $x = h = \dfrac{-b}{2a}$ 垂直线对称的图形.

Example 1.6

Find a quadratic equation with rational coefficients having $2-4\sqrt{3}$ as one of its roots.

Solution

Since $2-4\sqrt{3}$ is one root, so the other root must be $2+4\sqrt{3}$.
$$[x-(2-4\sqrt{3})][x-(2+4\sqrt{3})] = [(x-2)+4\sqrt{3}][(x-2)-4\sqrt{3}] = (x-2)^2 - (4\sqrt{3})^2$$
$$= x^2 - 4x - 44 = 0.$$

Example 1.7

The graph of $y = 3(x-4)(x+2)$ has a vertex P, find the coordinate of P.

Solution

Method 1 (By Symmetry)

The graph intersect x-axis at $x = 4, -2$; it is symmetry about the line $x = k$, where k is the x-coordinate of the vertex. Hence, the x-coordinate of the vertex would be in the middle of these two values, $x = \dfrac{4-2}{2} = 1$; $y = 3(1-4)(1+2) = -27$. The vertex $P(1, -27)$.

Method 2 (By Completing the Square)
$$y = 3(x-4)(x+2) = 3(x^2 - 2x - 8) = 3(x^2 - 2x) - 24 = 3(x^2 - 2x + 1 - 1) - 24$$
$$= 3(x^2 - 2x + 1) - 27 = 3(x-1)^2 - 27.$$
The vertex $P(1, -27)$.

Example 1.8

Find the minimum value of $x^2 - 6x + 5$ and the corresponding value of x.

Solution

$x^2 - 6x + 5 = x^2 - 6x + 9 - 4 = (x-3)^2 - 4$. The minimum value is -4 occurs at $x = 3$.

Example 1.9

Find all values of x such that $x^2 + 13x + 3$ is a perfect integral square.

Solution

Suppose $x^2 + 13x + 3 = k^2$, where k is an integer.
$$x^2 + 13x + 3 = k^2 \Rightarrow x^2 + 13x + \dfrac{13^2}{2^2} - \dfrac{13^2}{2^2} + 3 = k^2 \Rightarrow \left(x + \dfrac{13}{2}\right)^2 = k^2 + \dfrac{157}{4} \Rightarrow (2x+13)^2 = 4k^2 + 157$$
$$\Rightarrow (2x+13)^2 - 4k^2 = 157 \Rightarrow (2x+13+2k)(2x+13-2k) = 157.$$

Since 157 is a prime number, so we have 4 possibilities:
(1) $2x + 13 + 2k = 1, 2x + 13 - 2k = 157 \Rightarrow x = 33, k = -39.$
(2) $2x + 13 + 2k = 157, 2x + 13 - 2k = 1 \Rightarrow x = 33, k = 39.$

(3) $2x + 13 + 2k = -1$, $2x + 13 - 2k = -157 \Rightarrow x = -46$, $k = 39$.

(4) $2x + 13 + 2k = -157$, $2x + 13 - 2k = -1 \Rightarrow x = -46$, $k = -39$.

The values of x are 33 and -46.

Example 1.10

Determine all non-zero values of t such that all roots of $t(x-1)(x-2) = x$ are real.

Solution

$t(x-1)(x-2) = x \Rightarrow tx^2 - 3tx + 2t = x \Rightarrow tx^2 + (-3t-1)x + 2t = 0$.

We want all real roots, so we need $D \geq 0$.

$D = (-3t-1)^2 - 4(t)(2t) \geq 0 \Rightarrow t^2 + 6t + 1 \geq 0$.

When $t^2 + 6t + 1 = 0 \Rightarrow t = \dfrac{-6 \pm \sqrt{36-4}}{2} = -3 \pm 2\sqrt{2}$.

Hence, we need $t \geq -3 + 2\sqrt{2}$ or $t \leq -3 - 2\sqrt{2}$.

Example 1.11

What is the minimum value of $\dfrac{x^2 - 2x - 1}{(x-4)^2}$?

Solution

Suppose the minimum value exists and denoted by M.

$\dfrac{x^2 - 2x - 1}{(x-4)^2} = M \Rightarrow x^2 - 2x - 1 = M(x^2 - 8x + 16) \Rightarrow (M-1)x^2 + (2-8M)x + (1+16M) = 0$.

If M exists, it means x must be a real number; therefore, the discriminant $D \geq 0$.

$D = (2-8M)^2 - 4(M-1)(1+16M) \geq 0 \Rightarrow 28M \geq -8 \Rightarrow M \geq \dfrac{-2}{7}$.

The minimum value of $\dfrac{x^2 - 2x - 1}{(x-4)^2}$ is $\dfrac{-2}{7}$.

Example 1.12

A parabola has vertex $(6, 5)$ and axis of symmetry parallel to the y-axis. One of the x-intercept is 2, what is the value of the other x-intercept?

Solution

Let the other x-intercept be a, $\dfrac{a+2}{2} = 6 \Rightarrow a = 10$.

Example 1.13

Find the roots of $x^2 + \left(a - \dfrac{1}{a}\right)x - 1 = 0$.

Solution 解

$x^2 + \left(a - \dfrac{1}{a}\right)x - 1 = (x + a)\left(x - \dfrac{1}{a}\right) = 0 \Rightarrow x = -a \text{ or } \dfrac{1}{a}.$

We can also use quadratic formula to solve for the roots.

$x = \dfrac{\dfrac{1}{a} - a \pm \sqrt{\left(a - \dfrac{1}{a}\right)^2 + 4}}{2} = \dfrac{\dfrac{1}{a} - a \pm \sqrt{a^2 + 2 + \dfrac{1}{a^2}}}{2}$

$= \dfrac{\dfrac{1}{a} - a \pm \sqrt{\left(a + \dfrac{1}{a}\right)^2}}{2} = -a \text{ or } \dfrac{1}{a}.$

Example 例 1.14

If $(-2, 7)$ is the maximum point for the graph $y = -2x^2 - 4ax + k$. What is the value of k?

Solution 解

$y = -2x^2 - 4ax + k = -2(x^2 + 2ax) + k = -2(x^2 + 2ax + a^2 - a^2) + k = -2(x + a)^2 + 2a^2 + k.$
Clearly, $a = 2$, and $2a^2 + k = 7 \Rightarrow 2(2)^2 + k = 7 \Rightarrow k = -1.$

1.4 Solving Quadratic Inequality 解二次不等式

It is easier to solve the quadratic inequality by sketching the graph. We will work through with a few examples to see how it works.
用画图法求解二次不等式更加容易,可以通过一些例子来看是如何操作的.

Example 例 1.15

Solve $2x^2 < x + 3$.

Solution 解

$2x^2 - x - 3 < 0 \Rightarrow (2x - 3)(x + 1) < 0.$
The sketch of the graph is shown (Fig. 1.2).
The solution to this inequality would be $-1 < x < 1.5$.

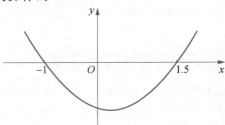

Fig. 1.2

Example 例 1.16

Solve the inequality $\dfrac{x^2}{x - 2} < x + 1, x \neq 2.$

Solution

In order to avoid the change sign of the inequality, we rearrange the inequality and multiply by $(x-2)^2$.

$$\frac{x^2}{x-2} - (x+1) < 0 \Rightarrow \frac{x^2 - (x-2)(x+1)}{x-2} < 0 \Rightarrow (x-2)^2 \frac{x^2 - (x-2)(x+1)}{x-2} < 0$$

$$\Rightarrow (x-2)[x^2 - (x+1)(x-2)] < 0 \Rightarrow (x-2)(x^2 - x^2 + x + 2) < 0$$

$$\Rightarrow (x-2)(x+2) < 0.$$

The solution is $-2 < x < 2$.

Example 1.17

Solve the inequality $\frac{2x-4}{x} \geqslant 7$.

Solution

We rearrange the inequality; then multiply x^2 to both sides of the inequality.

$$\frac{2x-4}{x} - 7 \geqslant 0 \Rightarrow \frac{2x - 4 - 7x}{x} \geqslant 0 \Rightarrow -x(5x+4) \geqslant 0.$$

The solution to the inequality is $\frac{-4}{5} \leqslant x < 0$, we excluded 0 because x can't equal to 0.

Example 1.18

Find the values of m for the equation $x^2 - 4x + 2 = m$ such that
(a) two distinct real roots; (b) repeated roots; (c) no real roots (complex roots).

Solution

Use Discriminant $D = b^2 - 4ac$.

Before using discriminant, we need to rewrite the equation as $x^2 - 4x + 2 - m = 0$.

$D = (-4)^2 - 4 \cdot 1 \cdot (2-m) = 16 - 8 + 4m = 8 + 4m$.

(a) Two distinct real roots, we need $8 + 4m > 0 \Rightarrow m > -2$.
(b) repeated roots, we need $8 + 4m = 0 \Rightarrow m = -2$.
(c) No real roots, we need $8 + 4m < 0 \Rightarrow m < -2$.

1.5 Intersection between Line and Parabola
直线与抛物线相交

We consider a quadratic curve with equation $y = ax^2 + bx + c$ and a straight line $y = mx + n$. The intersection points exist when the values of x and y satisfy both the equations. This means, in order to find the intersection points we solve the equation $ax^2 + bx + c = mx + n \Rightarrow ax^2 + (b-m)x + (c-n) = 0$. The discriminant

$D = (b-m)^2 - 4a(c-n)$ can tell us what kind of situations we have.

假设现在有二次曲线方程 $y = ax^2 + bx + c$ 和直线方程 $y = mx + n$. 当 x 和 y 的值同时满足两个方程时, 说明两个图像相交. 也就是说, 如果要找到相交的点, 就要先解以下这个方程: $ax^2 + bx + c = mx + n \Rightarrow ax^2 + (b-m)x + (c-n) = 0$. $D = (b-m)^2 - 4a(c-n)$ 这个判别式能帮助我们判断具体是哪种情况.

Fig. 1.3

When $D > 0$, it means the line and the parabola intersect at two distinct points (Fig. 1.3).

当 $D>0$, 说明直线与抛物线有两个不同的交点(图1.3).

When $D < 0$, it means the line and the parabola do not intersect (Fig. 1.4).

当 $D<0$, 说明直线与抛物线不相交(图1.4).

When $D = 0$, it means the line and the parabola intersect at one point; in this case we said the line is a tangent line to the parabola (Fig. 1.5).

当 $D=0$, 说明直线与抛物线只有唯一一个交点, 称该直线是抛物线的切线(图1.5).

Fig. 1.4

Fig. 1.5

Example 例 1.19

A straight line $y = kx + 1$ and the curve $y = x^2 - 7x + 2$, find the values of k such that

(a) the line is a tangent to the curve; (b) intersect at two distinct points; (c) no intersections.

Solution 解

$x^2 - 7x + 2 = kx + 1 \Rightarrow x^2 - (7+k)x + 1 = 0$.

$D = (7+k)^2 - 4$.

(a) The line is a tangent to the curve, so $D = 0 \Rightarrow (7+k)^2 - 4 = 0 \Rightarrow 7+k = \pm 2 \Rightarrow k = -5$ or -9.

(b) The line intersects the curve at two distinct points, so $D > 0$.

$(7+k)^2 - 4 > 0 \Rightarrow k > -5$ or $k < -9$.

(c) The line and the curve have no intersections, so $D < 0$.

$(7+k)^2 - 4 < 0 \Rightarrow -9 < k < -5$.

Example 例 1.20

The line $y = k - 3x$ is a tangent to the curve $x^2 + 2xy - 20 = 0$, find the value of k and the intersection point.

Solution 解

$x^2 + 2xy - 20 = 0 \Rightarrow y = \dfrac{20 - x^2}{2x}$; $\dfrac{20 - x^2}{2x} = k - 3x \Rightarrow 20 - x^2 = 2x(k - 3x) \Rightarrow 5x^2 - 2kx + 20 = 0$.

$D = 4k^2 - 400 = 0 \Rightarrow k = \pm 10$.

When $k = 10 \Rightarrow 5x^2 - 20x + 20 = 0 \Rightarrow x^2 - 4x + 4 = (x - 2)^2 = 0 \Rightarrow x = 2$, $y = 10 - 3(2) = 4$.

The intersection point is $(2, 4)$.

When $k = -10 \Rightarrow 5x^2 + 20x + 20 = 0 \Rightarrow x^2 + 4x + 4 = (x + 2)^2 = 0 \Rightarrow x = -2$, $y = -10 - 3(-2) = -4$.

The intersection point is $(-2, -4)$.

Example 1.21

A curve has equation $y = -\dfrac{1}{x} + 2k$ and a line has equation $y = kx + 4$, where k is a constant.

(a) Find the set of values of k for which the curve and the line meet at two distinct points.

(b) The line is a tangent to the curve for two particular values of k. For each of these values find the coordinate of the point at which the tangent touches the curve.

Solution

(a) $-\dfrac{1}{x} + 2k = kx + 4 \Rightarrow \dfrac{-1 + 2kx}{x} = kx + 4 \Rightarrow -1 + 2kx = x(kx + 4) \Rightarrow kx^2 + (4 - 2k)x + 1 = 0$.

$D = (4 - 2k)^2 - 4k > 0 \Rightarrow 4k^2 - 20k + 16 > 0 \Rightarrow k > 4$ or $k < 1$.

(b) The line is tangent to the curve when $k = 4$ or $k = 1$.

When $k = 4 \Rightarrow 4x^2 - 4x + 1 = 0 \Rightarrow (2x - 1)^2 = 0 \Rightarrow x = \dfrac{1}{2}$.

When $k = 1 \Rightarrow x^2 + 2x + 1 = 0 \Rightarrow (x + 1)^2 = 0 \Rightarrow x = -1$.

Example 1.22

The equation of a curve is $y = -2x^2 - 4x + c$, where c is a constant.

(a) Find the set of values of c for which the whole of the curve lies below the x-axis.

(b) Find the values of c for which the line $y - 4x = 3$ is a tangent to the curve.

Solution

(a) $y = -2x^2 - 4x + c = -2x^2 - 4x - 2 + 2 + c = -2(x^2 + 2x + 1) + 2 + c = -2(x + 1)^2 + 2 + c$.

Since $-2(x + 1)^2$ is always smaller than or equal to zero.

In order for the curve lies below the x-axis, we need $2 + c < 0 \Rightarrow c < -2$.

(b) $-2x^2 - 4x + c = 4x + 3 \Rightarrow 2x^2 + 8x + 3 - c = 0$.

We need $\Delta = 8^2 - 4 \times 2 \times (3 - c) = 0 \Rightarrow c = -5$.

Example 1.23

Express $2x^2 - 8x + 15$ in the form $a(x + b)^2 + c$, where a, b and c are constants.

Solution

$2x^2 - 8x + 15 = 2(x^2 - 4x) + 15 = 2(x^2 - 4x + 4 - 4) + 15$
$= 2(x^2 - 4x + 4) - 8 + 15 = 2(x - 2)^2 + 7$.

Example 1.24

Find the set of values of c for which the equation $3x^2 + 4cx + 2c = 0$ has no real roots.

Solution

We want $D = (4c)^2 - 4 \times 3 \times 2c = 16c^2 - 24c < 0 \Rightarrow 0 < c < \dfrac{3}{2}$.

Example 1.25

Find the x-coordinates of the points of intersection of the curve $y = x^{\frac{1}{2}} - 3$ with the curve $y = 3x^{\frac{1}{4}} - 5$.

Solution

We want to solve $x^{\frac{1}{2}} - 3 = 3x^{\frac{1}{4}} - 5$.

Let $x^{\frac{1}{4}} = u \Rightarrow u^2 - 3 - 3u + 5 = 0 \Rightarrow u^2 - 3u + 2 = 0 \Rightarrow (u-1)(u-2) = 0$.

$u = 1$ or $u = 2 \Rightarrow x^{\frac{1}{4}} = 1$ or $x^{\frac{1}{4}} = 2 \Rightarrow x = 1$ or $x = 16$.

Example 1.26

Find the set of values of k for which the curve $y = -\dfrac{3}{x}$ and the line $y = kx + 3$ meet at two distinct points.

Solution

$-\dfrac{3}{x} = kx + 3 \Rightarrow kx^2 + 3x + 3 = 0$.

We want $\Delta = 3^2 - 4 \times k \times 3 > 0 \Rightarrow 9 - 12k > 0 \Rightarrow k < \dfrac{3}{4}$.

Example 1.27

A farmer wants to add an extension with a floor to the back of his house. He uses 24 m of fencing. The dimensions, in meters, are coordinates x and y as shown (Fig. 1.6).

(a) Show that the area of floor, A m^2, is given by $A = 24x - 2x^2$.

(b) Given that x and y can vary, find the dimensions of floor for which the value of A is a maximum.

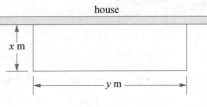

Fig. 1.6

Solution

(a) $2x + y = 24 \Rightarrow y = 24 - 2x$.

Area $= xy = x(24 - 2x) = 24x - 2x^2$.

(b) $A = 24x - 2x^2 = -2(x^2 - 12x) = -2(x^2 - 12x + 36) + 72 = -2(x-6)^2 + 72$.

In order to get maximum area, we need $x = 6 \Rightarrow y = 24 - 2 \times 6 = 12$.

The dimension is 6 m $\times$ 12 m.

Example 1.28

(a) Express $3x^2 + 6x + 15$ in the form $a(x+b)^2 + c$, where a, b and c are constants.

(b) Find the set of values of x for which $3x^2 + 6x + 15 < 39$.

Solution

(a) $3x^2 + 6x + 15 = 3(x^2 + 2x + 1) - 3 + 15 = 3(x+1)^2 + 12$.

(b) $3x^2 + 6x + 15 < 39 \Rightarrow 3x^2 + 6x - 24 < 0 \Rightarrow x^2 + 2x - 8 < 8 \Rightarrow (x-2)(x+4) < 0$.
So $-4 < x < 2$.

Example 1.29

A curve has equation $y = 3x^2 + 2x - 1$.

(a) Find the set of values of x for which $y > 0$.

(b) Find the value of c for which the line $y = 8x + 2c$ is a tangent to the curve.

Solution

(a) $y = 3x^2 + 2x - 1 > 0 \Rightarrow (3x-1)(x+1) > 0 \Rightarrow x > \dfrac{1}{3}$ or $x < -1$.

(a) $3x^2 + 2x - 1 = 8x + 2c \Rightarrow 3x^2 - 6x + (2c - 1) = 0$.
We want $\Delta = (-6)^2 - 4 \times 3 \times (2c - 1) = 36 - 24c + 12 = 48 - 24c = 0 \Rightarrow c = 2$.

Example 1.30

Find the set of values of a for which the curve $y = 2x^2 - ax$ and the line $y = ax - 2$ meet at two distinct points.

Solution

$2x^2 - ax = ax - 2 \Rightarrow 2x^2 - 2ax + 2 = 0 \Rightarrow x^2 - ax + 1 = 0$.
We want $D = a^2 - 4 \geqslant 0 \Rightarrow 2 \leqslant a$ or $a \leqslant -2$.

*1.6 Extended Material 拓展内容

Consider a quadratic equation $y = ax^2 + bx + c = a\left(x^2 + \dfrac{b}{a}x + \dfrac{c}{a}\right)$, if it has two linear factors namely $(x - \alpha)$ and $(x - \beta)$.

$a(x - \alpha)(x - \beta) = a[x^2 - (\alpha + \beta)x + \alpha\beta] = a\left(x^2 + \dfrac{b}{a}x + \dfrac{c}{a}\right)$
$= a[x^2 - (\text{sum of roots})x + \text{product of roots}]$.

So $\alpha + \beta = \dfrac{-b}{a}$ and $\alpha\beta = \dfrac{c}{a}$. That means the sum of the roots of a quadratic equation $ax^2 + bx + c = 0$ is $-\dfrac{b}{a}$, and the product of roots is $\dfrac{c}{a}$. The question we might have now is what is $\alpha^n + \beta^n$, $n \in \mathbf{Z}$?

Consider the case where $n = 2$,

$$\alpha^2 + \beta^2 = (\alpha + \beta)^2 - 2\alpha\beta = \left(\dfrac{-b}{a}\right)^2 - 2\dfrac{c}{a} = \dfrac{b^2 - 2ac}{a^2}.$$

When $n = -1$,

$$\dfrac{1}{\alpha} + \dfrac{1}{\beta} = \dfrac{\alpha + \beta}{\alpha\beta} = \dfrac{\tfrac{-b}{a}}{\tfrac{c}{a}} = \dfrac{-b}{c}.$$

Notice that $\alpha + \beta$, $\alpha^2 + \beta^2$ and $\dfrac{1}{\alpha} + \dfrac{1}{\beta}$ are symmetric function of α and β; that is, if we interchange α and β, the expression remains unchanged.

Example 例 1.31

The quadratic equation $x^2 + 5x + 7 = 0$ has roots α and β. Find an equation with roots 2α and 2β.

Solution 解

Method 1

$\alpha + \beta = -5$, $\alpha\beta = 7$. So $2\alpha + 2\beta = 2(\alpha + \beta) = -10$ and $2\alpha \cdot 2\beta = 4\alpha\beta = 28$; therefore $x^2 + 10x + 28 = 0$ has roots 2α and 2β.

Method 2

Let $u = 2\alpha$, $v = 2\beta$; so $\alpha = \dfrac{u}{2}$, $\beta = \dfrac{v}{2}$.

Since α and β are roots for $x^2 + 5x + 7 = 0 \Rightarrow \left(\dfrac{u}{2}\right)^2 + 5\left(\dfrac{u}{2}\right) + 7 = 0 \Rightarrow \dfrac{u^2}{4} + \dfrac{5u}{2} + 7 = 0 \Rightarrow u^2 + 10u + 28 = 0$. ∎

Example 例 1.32

Given an equation $ax^2 + bx + c = 0$, one root is twice of the other root. Prove $2b^2 = 9ac$.

Solution 解

Let the two roots be α and 2α. So $\alpha + 2\alpha = \dfrac{-b}{a} \Rightarrow \alpha = \dfrac{-b}{3a}$. Also $\alpha \cdot 2\alpha = \dfrac{c}{a} \Rightarrow 2\left(\dfrac{-b}{3a}\right)^2 = \dfrac{c}{a} \Rightarrow \dfrac{2b^2}{9a^2} = \dfrac{c}{a} \Rightarrow 2b^2 = \dfrac{9a^2 c}{a} = 9ac$. ∎

1.7 Recursive Formula for the nth Power of Roots
根的 n 次幂之和的递归公式

Consider the general quadratic equation $ax^2 + bx + c = 0$ with roots α and β. Let $S_n = \alpha^n + \beta^n$, the sum of the nth power of the roots. Since we know $a\alpha^2 + b\alpha + c = 0$ and $a\beta^2 + b\beta + c = 0$; we now add these two equations we would obtain:

$$a(\alpha^2 + \beta^2) + b(\alpha + \beta) + 2c = 0 \Rightarrow aS_2 + bS_1 + 2c = 0 \Rightarrow S_2 = \frac{-2c - bS_1}{a} = \frac{b^2 - 2ac}{a^2}.$$

Similarly, we can have $a\alpha^{2+n} + b\alpha^{1+n} + c\alpha^n = 0$ and $a\beta^{2+n} + b\beta^{1+n} + c\beta^n = 0$ which could be obtained by multiplying α^n and β^n to $a\alpha^2 + b\alpha + c = 0$ and $a\beta^2 + b\beta + c = 0$ respectively.

This implies $a(\alpha^{2+n} + \beta^{2+n}) + b(\alpha^{1+n} + \beta^{1+n}) + c(\alpha^n + \beta^n) = 0$ which is equivalent to $aS_{2+n} + bS_{1+n} + cS_n = 0$.

Example 例 1.33

Suppose the equation $x^2 + 2x - 2$ has two roots α and β. Find the values of $\alpha^3 + \beta^3$, $\alpha^4 + \beta^4$ and $\alpha^5 + \beta^5$.

Solution 解

Let $S_n = \alpha^n + \beta^n \Rightarrow S_{n+2} + 2S_{n+1} - 2S_n = 0$. We know $S_1 = -2$ and $S_2 = 8$

$\Rightarrow S_3 = 2S_1 - 2S_2 = -4 - 16 = -20;$
$\Rightarrow S_4 = 2S_2 - 2S_3 = 16 + 40 = 56;$
$\Rightarrow S_5 = 2S_3 - 2S_4 = -40 - 112 = -152.$

Summary of Key Theories 核心定义总结

(1) A quadratic equation is an equation written in the form of $ax^2 + bx + c$, where a, b and c are constants and $a \neq 0$.

(2) The root of a quadratic equation $ax^2 + bx + c$ is the value $x = \alpha$ such that $a\alpha^2 + b\alpha + c = 0$.

(3) A quadratic equation $ax^2 + bx + c = 0$ where $a \neq 0$, then the roots of the equation are $x = \dfrac{-b \pm \sqrt{b^2 - 4ac}}{2a}$.

Functions

第 2 章 函 数

As used in ordinary language, the word **function** indicates dependence of one quantity on another quantity. For example, if your math teacher tells you that your final grade in the math class will be a function of your exam scores, you interpret this to mean that the teacher has some rules for converting the scores in exams into final grade. More generally, suppose two sets of objects are given; the relationship between the elements of the sets is called a **function**.

在数学的常用语中,函数这个词的含义是一个量对于另一个量的依赖性.比如,如果数学老师告诉你,你的最终成绩是平时测评成绩的一个函数,那么你就该这么理解:数学老师会根据设定的规则,将测评成绩转变成最终成绩.更通俗地讲,假设有两组元素,这两组元素之间的关系便可称为函数.

Definition 定义

A ***function*** f from a set X to a set Y is a rule that assigns a unique element $f(x)$ in Y (outputs) to a unique element x in X (input), with the property that each input is related to only one output. An alternative name for a function is a **mapping**.

The notation $f: X \to Y$ means f is a function from X to Y, where X is called the ***domain*** of f, and the set of all values of f taken together is called the **range** of f.

Consider a function $y = f(x)$, then the **independent variable** is x, and y is the **dependent variable**.

从一个集合 X 到一个集合 Y 的函数是一个规则,它将 Y(输出)中的一个唯一元素 $f(x)$ 赋给 X(输入)中的每个元素 x,特点是每个输入只与一个输出相关.因此,函数的另一个名称是映射.

$f: X \to Y$ 这个表达是指 f 是从 X 到 Y 的一个函数, X 被称为 Y 的定义域, f 所有的值的集合被称为 f 的值域.

假设一个函数表达式为 $y = f(x)$,那么,它的自变量是 x,因变量是 y.

Another way to think of a function is as a machine. Suppose f is a function from X to Y and an input x of X is given. Imagine f to be a machine that processes x in a certain way to produce the output $f(x)$. This is demonstrated in Fig. 2.1.

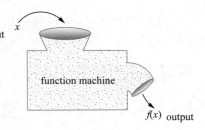
Fig. 2.1

Example 例 2.1

Find the domain and range for the following functions.

(a) $f(x) = x^2 + x + 2$; (b) $f(x) = -2x^2 - 4x - 3$; (c) $y = \dfrac{1}{x}$.

Solution 解

(a) We know the graph of the function is a parabola opens upwards.

$f(x) = x^2 + x + 2 = x^2 + x + \dfrac{1}{4} + \dfrac{7}{4} = \left(x + \dfrac{1}{2}\right)^2 + \dfrac{7}{4}$; the minimum point is $\left(-\dfrac{1}{2}, \dfrac{7}{4}\right)$.

Hence, the domain of $f(x)$ is the set of all real numbers; we denoted it by **R**, or we can use interval notation $(-\infty, \infty)$ or write it as inequality $-\infty < x < \infty$.

The range is $y \geqslant \dfrac{7}{4}$ or $\left[\dfrac{7}{4}, \infty\right)$.

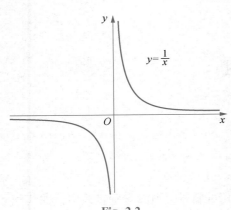
Fig. 2.2

(b) We know the graph of the function is a parabola opens downwards.
$f(x) = -2x^2 - 4x - 3 = -2(x^2 + 2x) - 5 = -2(x^2 + 2x + 1 - 1) - 5 = -2(x^2 + 2x + 1) + 2 - 5 = -2(x + 1)^2 - 3$; the maximum point is $(-1, -3)$.

Hence, the domain of $f(x)$ is **R**, and the range is $y \leqslant -3$.

(c) From IGCSE, we know the graph of the $f(x) = \dfrac{1}{x}$ is a rectangular hyperbola. It has a vertical asymptote at $x = 0$ and a horizontal asymptote at $y = 0$ (Fig. 2.2).

Hence, the domain is $x \neq 0$ and the range is $y \neq 0$. ∎

Example 例 2.2

Find the domain of the function $f(x) = \dfrac{x}{x^2 - 4}$.

Solution 解

$f(x) = \dfrac{x}{x^2 - 4} = \dfrac{x}{(x+2)(x-2)}$. The domain consists all real numbers except $x = \pm 2$.

We can also use set notation to represent domain.
Domain: $x \neq \pm 2$ or $(-\infty, -2) \cup (-2, 2) \cup (2, \infty)$. ∎

Example 例 2.3

Find the range of the following function with the specified domain.

(a) $f(x) = x^2 - x + 3$ for $x \in [-2, 5]$; (b) $g(x) = -x^2 + x - 5$ for $0 \leq x \leq 6$;
(c) $h(x) = x^2 - 2x + 3$ for $-2 < x \leq 3$; (d) $i(x) = -x^2 + 2x - 5$ for $x \in (2, 5)$.

Solution 解

(a) $f(x) = x^2 - x + 3 = \left(x - \frac{1}{2}\right)^2 + \frac{11}{4}$, the minimum point is $\left(\frac{1}{2}, \frac{11}{4}\right)$ and $x = \frac{1}{2}$ is in the given domain. $f(-2) = 9$, $f(5) = 23$. Hence, the range is $\frac{11}{4} \leq y \leq 23$.

(b) $g(x) = -x^2 + x - 5 = -\left(x - \frac{1}{2}\right)^2 - \frac{19}{4}$, the maximum point is $\left(\frac{1}{2}, \frac{-19}{4}\right)$ and $x = \frac{1}{2}$ is in the given domain. $g(0) = -5$, $g(6) = -35$. Hence, the range is $-35 \leq y \leq \frac{-19}{4}$.

(c) $h(x) = x^2 - 2x + 3 = (x - 1)^2 + 2$, the minimum point is $(1, 2)$ and $x = 1$ is in the given domain. $h(-2) = 11$, $h(3) = 6$. Hence, the range is $2 \leq y < 11$.

(d) $i(x) = -x^2 + 2x - 5 = -(x - 1)^2 - 4$, the maximum point is $(1, -4)$ and $x = 1$ is not in the given domain. $i(2) = -5$, $i(5) = -20$. Hence, the range is $-20 < y < -5$.

2.1 The Vertical Line Test 垂直线测试

Given a curve in the xy-plane, due to the property that each input is related to only one output, if it represents a graph of a function $f(x)$, then no vertical line intersects the curve more than once.
The following Figures illustrate how to identity whether a curve is a function or not.
已知 xy 平面上有一条曲线，根据函数每个输入值仅对应一个输出值的特性，则没有垂直线与该曲线相交超过 1 次以上. 图 2.3 说明如何辨识曲线是否为函数.

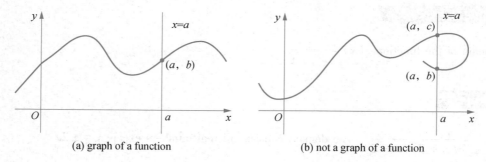

(a) graph of a function   (b) not a graph of a function

Fig. 2.3

Example 例 2.4

Determine which of the following graph (Fig. 2.4) represents a graph of a function $y = f(x)$.

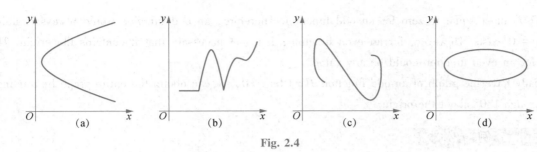

(a)　　　　　　(b)　　　　　　(c)　　　　　　(d)

Fig. 2.4

 解

Using the vertical line test, it is clear that (a), (c) and (d) failed. So they can not be graphs for functions. (b) is the only graph that is a graph of a function.

2.2 Symmetry of Functions　函数的对称性

Definition　定义

A function $f(x)$ is called an even function if it satisfies $f(-x) = f(x)$ $\forall x$ in the domain of $f(x)$ ($\forall$ is the mathematical symbol means "for all"). The graph of an even function is symmetry about the y-axis [Fig.2.5(a)].

A function $f(x)$ is called an odd function is it satisfies $f(-x) = -f(x)$ $\forall x$ in the domain of $f(x)$. The graph of an odd function is symmetry about the origin [Fig.2.5(b)].

如果对于函数 $f(x)$ 定义域内的任意一个 x 满足 $f(-x) = f(x)$，则该函数为偶函数，偶函数的图像关于 y 轴对称，如图 2.5(a) 所示.

如果对于函数 $f(x)$ 定义域内的任意一个 x 满足 $f(-x) = -f(x)$，则该函数为奇函数，奇函数的图像关于原点对称，如图 2.5(b) 所示.

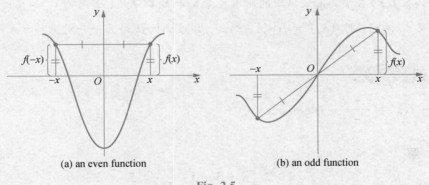

(a) an even function　　　　　　(b) an odd function

Fig. 2.5

Due to the definition of an odd function, if $x = 0$ is in the domain of $f(x)$.

Since $f(-x) = -f(x) \Rightarrow f(-0) = -f(0) \Rightarrow f(0) = -f(0)$.

Hence, $f(0)$ must equal to zero for an odd function. Therefore, an odd function would always contain the origin if $x = 0$ exists. However, for an even function; it is not necessary that it contains the origin. The y-intercept for an even function could be any value.

If we already have the graph of an odd function $f(x)$ for $x>0$, we can obtain the entire graph by rotating this portion through $180°$ about the origin.

Example 例 2.5

Determine whether each of following functions is even, odd or neither.
(a) $f(x) = x^3 + 2x$; (b) $g(x) = 1 - 3x^4$; (c) $h(x) = 4x - 3x^2$.

Solution 解

We start with the definition of being an even/odd function.
(a) $f(-x) = (-x)^3 + 2(-x) = -x^3 - 2x = -(x^3 + 2x) = -f(x)$. So it is an odd function.
(b) $g(-x) = 1 - 3(-x)^4 = 1 - 3x^4 = g(x)$. So it is an even function.
(c) $h(-x) = 4(-x) - 3(-x)^2 = -4x - 3x^2 \neq h(x) \neq -h(x)$. So it is neither even nor odd. ∎

2.3 One-to-one Function 一一对应函数

Definition 定义

A function $f(x)$ is called a **one-to-one function** or **injective mapping** if, and only if, for all elements x_1 and x_2 in the domain:
If $f(x_1) = f(x_2)$, then $x_1 = x_2$.
Or, equivalently,
if $x_1 \neq x_2$, then $f(x_1) \neq f(x_2)$.
当且仅当一个函数$f(x)$对于其定义域中的所有元素x_1和x_2都满足,如果$f(x_1) = f(x_2)$,则$x_1 = x_2$;同理类推,如果$x_1 \neq x_2$,则$f(x_1) \neq f(x_2)$,那么,称该函数$f(x)$为一一对应函数或单射函数.

Example 例 2.6

Prove the function $f(x) = \dfrac{x+2}{x-3}$ is one-to-one.

Solution 解

We first simplify the expression. $f(x) = \dfrac{x+2}{x-3} = \dfrac{x-3+5}{x-3} = 1 + \dfrac{5}{x-3}$.

Suppose $f(x_1) = f(x_2) \Rightarrow 1 + \dfrac{5}{x_1 - 3} = 1 + \dfrac{5}{x_2 - 3} \Rightarrow \dfrac{5}{x_1 - 3} = \dfrac{5}{x_2 - 3} \Rightarrow (x_1 - 3) = (x_2 - 3) \Rightarrow x_1 = x_2$.

Hence, it is a one-to-one function.

Example 例 2.7

Show the function $f(x) = x^2 + 2x - 1$ is not one-to-one.

Solution 解

We first simplify the expression. $f(x) = x^2 + 2x - 1 = (x + 1)^2 - 2$.
Suppose $f(x_1) = f(x_2) \Rightarrow (x_1 + 1)^2 - 2 = (x_2 + 1)^2 - 2 \Rightarrow x_1 + 1 = \pm(x_2 + 1) \Rightarrow x_1 = x_2$ or $x_1 = -x_2 - 2$.
Hence, it is not a one-to-one function; because $f(1) = 2$ and $f(-3) = 2$, so we have two different values of x which would produce the same value of y.

2.4 The Horizontal Line Test 水平线测试

Given a graph of a function, in order to see if it is a one-to-one, we can use the horizontal line test. Due to the definition of being a one-to-one function, if the graph represents a one-to-one function $f(x)$, then no horizontal line intersects the curve more than once.
已知一个函数图像，为了检测它是否是一一对应函数，可以使用水平线测试.根据一一对应函数的定义，若该图像代表一一对应函数 $f(x)$，则任何水平线与该曲线相交点的个数不会多于一个.
The following figures illustrate how to identity whether a graph of function is one-to-one or not.

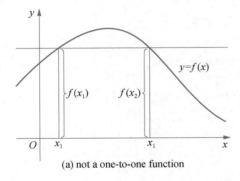

Fig. 2.6

2.5 Composite Function 复合函数

Consider two functions $f(x)$ and $g(x)$ and imagine that each is represented by a machine. If the two machines are hooked up so that the output from $f(x)$ is used as input to $g(x)$, then they work together to operate as one larger machine. Combining functions in this way is called composing them; the resulting function is called the composition of the two functions.

👑 Definition 定义

Let $f: X \to Y'$ and $g: Y \to Z$ be functions with the property that the range of f is a subset of the domain of g. We define a new function $g \circ f: X \to Z$ as $g \circ f(x) = g(f(x))$ $\forall x \in X$, where $g(f(x))$ is read "g of f of x". The function $g \circ f(x) = g(f(x))$ is called the composition of f and g.

假设有 $f: X \to Y'$ 和 $g: Y \to Z$ 两个函数，其中函数 f 的值域是函数 g 的定义域的子集. 可以由此推出 $g \circ f: X \to Z$ 的新函数为 $g \circ f(x) = g(f(x))$ $\forall x \in X$，其中 $g(f(x))$ 读作 "g of f of x". 该函数 $g \circ f(x) = g(f(x))$ 称为 f 和 g 的复合函数.

Note 注意

In CIE exam, $g \circ f(x)$ is written as $gf(x)$.

The definition is showed schematically in Fig. 2.7.

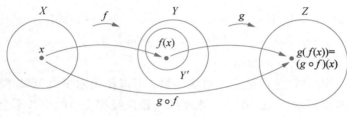

Fig. 2.7

📝 Example 例 2.8

Given $f(x) = (x-2)^2 + 1$ and $g(x) = \dfrac{3}{x-2}$. Find the followings:

(a) $fg(x)$;   (b) $gf(x)$;   (c) $ff(x)$;   (d) $gg(x)$.

Solution 解

(a) $fg(x) = \left(\dfrac{3}{x-2} - 2\right)^2 + 1$.

(b) $gf(x) = \dfrac{3}{(x-2)^2 + 1 - 2} = \dfrac{3}{(x-2)^2 - 1} = \dfrac{3}{(x-2+1)(x-2-1)} = \dfrac{3}{(x-1)(x-3)}$.

(c) $ff(x) = [(x-2)^2 + 1 - 2]^2 + 1 = [(x-2)^2 - 1]^2 + 1 = (x-3)^2(x-1)^2 + 1$.

(d) $gg(x) = \dfrac{3}{\dfrac{3}{x-2} - 2} = \dfrac{3}{\dfrac{3 - 2x + 4}{x-2}} = \dfrac{3x - 6}{7 - 2x}$.

Note 注意

In general, $fg(x) \neq gf(x)$.

Example 例 2.9

Given that $f(x) = g(x+1)$ and $g(x) = x^2$, what is $f(5)$?

Solution 解

$f(5) = g(5+1) = g(6) = 6^2 = 36$.

Example 例 2.10

Given $f(x-2) = x^3 - 2x + 1$, find $f(0)$.

Solution 解

$f(0) = f(2-2) = 2^3 - 2 \cdot 2 + 1 = 5$.

Example 例 2.11

If $f(x+1) = \dfrac{2f(x) + 1}{4}$ and $f(2) = 1$, find the value of $f(1)$.

Solution 解

$f(x+1) = \dfrac{2f(x)+1}{4} \Rightarrow f(2) = f(1+1) = \dfrac{2f(1)+1}{4} = 1 \Rightarrow f(1) = \dfrac{3}{2}$.

Example 例 2.12

Given $f(x) = x^2 - 2x$, find the sum of the values of x such that $f(x) = ff(x)$.

Solution 解

$ff(x) = (x^2 - 2x)^2 - 2(x^2 - 2x)$.
$f(x) = ff(x) \Rightarrow x^2 - 2x = (x^2 - 2x)^2 - 2(x^2 - 2x) \Rightarrow x^4 - 4x^3 + 2x^2 + 4x = 0$.
The values of x satisfy $f(x) = ff(x)$ would be the roots to the equation $x^4 - 4x^3 + 2x^2 + 4x = 0$. Hence, the sum of the roots is 4.

Note 注意

Given $f(x) = a_0 + a_1 x + a_2 x^2 + \cdots + a_n x^n$, the sum of the roots will always be $\dfrac{-a_{n-1}}{a_n}$.

Example 例 2.13

Let $f(x) = 3x^2 - 2x + 5$ and $fg(x) = 12x^4 + 56x^2 + 70$, find all possible values for the sum of the coefficients of $g(x)$.

Solution 解

The composite function $fg(x)$ is a polynomial; therefore, $g(x)$ is also a polynomial.
The sum of coefficients of a polynomial $g(x)$ would be $g(1)$.
$fg(1) = 12 + 56 + 70 = 138 \Rightarrow 3[g(1)]^2 - 2(g(1)) + 5 = 138 \Rightarrow [g(1) - 7][3g(1) + 19] = 0 \Rightarrow g(1) =$

7 or $\dfrac{-19}{3}$.

Example 例 2.14

A function f is defined for integers x and y as follows:
$f(xy) = f(x) \cdot f(y) - f(x+y) + 2019$, where either x or y is 1, and $f(1) = 2$.
(a) Prove that $f(n) = f(n-1) + 2019$; (b) Determine the value of $f(2019)$.

Solution 解

(a) Without Lose of Generality, let $x = 1$, $y = n - 1$.
$$f(xy) = f(x) \cdot f(y) - f(x+y) + 2019 \Rightarrow f(n-1) = f(1) \cdot f(n-1) - f(n) + 2019$$
$$\Rightarrow f(n-1) = 2 \cdot f(n-1) - f(n) + 2019 \Rightarrow f(n) = f(n-1) + 2019.$$
(b) $f(2019) = f(2018) + 2019 = f(2017) + 2 \times 2019 = \cdots = f(1) + 2018 \times 2019 = 4074344$.

Example 例 2.15

Given a function $f(x)$ satisfies the relation $f(x) + x \cdot f(1-x) = 1 + x^2$, $\forall x \in \mathbf{R}$. Find $f(x)$.

Solution 解

By observing the relation, we have a term $1 - x$ appears. Since the relation holds for all real values of x, so we now replace x by $1 - x$ in the equation; this implies $f(1-x) + (1-x) \cdot f(x) = 1 + (1-x)^2$.
We now have two equations
$$f(x) + x \cdot f(1-x) = 1 + x^2, \tag{1}$$
$$f(1-x) + (1-x) \cdot f(x) = 1 + (1-x)^2. \tag{2}$$
To eliminate $f(1-x)$, we multiply equation (2) by x and subtract equation (1)
$\Rightarrow (1-x)x \cdot f(x) - f(x) = x[1 + (1-x)^2] - 1 - x^2 \Rightarrow (x - x^2 - 1) \cdot f(x) = x^3 - 3x^2 + 2x - 1$
$\Rightarrow f(x) = \dfrac{x^3 - 3x^2 + 2x - 1}{(x - x^2 - 1)}$.

2.6　Inverse Function　反函数

Definition 定义

If $f(x)$ is one-to-one, then it has as inverse function $f^{-1}(x)$. The value of $f^{-1}(x)$ is the unique number y in the domain of $f(x)$ such that $f(y) = x$. It means $y = f^{-1}(x)$ if and only if $x = f(y)$.
如果函数$f(x)$是一一对应函数，那么，它的反函数为$f^{-1}(x)$. $f^{-1}(x)$的值是唯一的数值y，y是$f(x)$的定义域，所以，$f(y) = x$. 这表示当且仅当$x = f(y)$，$y = f^{-1}(x)$.

Consider a function $f(x) = x^3$, it is easy to show that it is a one-to-one.
$y = x^3$ is equivalent to $x = y^{\frac{1}{3}}$, we now reversing the roles of x and y.
$y = x^{\frac{1}{3}} \Leftrightarrow x = y^3$. The inverse function of $f(x) = x^3$ is therefore $f^{-1}(x) = x^{\frac{1}{3}}$.
The steps for finding an inverse function for a one-to-one function $f(x)$ are:
(1) Write $y = f(x)$.
(2) Interchange x and y.
(3) Make y the subject, the resulting equation is $y = f^{-1}(x)$.

Example 例 2.16

Show $f(x) = 4x + 2$ is a one-to-one, and then find $f^{-1}(x)$.

Solution 解

Suppose $f(x_1) = f(x_2) \Rightarrow 4x_1 + 2 = 4x_2 + 2 \Rightarrow x_1 = x_2$; hence it is one-to-one.
$f^{-1}(x): x = 4y + 2 \Rightarrow y = \dfrac{x-2}{4}$. Hence, $f^{-1}(x) = \dfrac{x-2}{4}$. ∎

There are several things we should know about the relationship between $f(x)$ and $f^{-1}(x)$.
If $y = f^{-1}(x) \Leftrightarrow x = f(y)$, this implies that the domain of $f^{-1}(x)$ is the range of $f(x)$ and vice versa.
The inverse of a one-to-one function is itself one-to-one, therefore it also has an inverse.
The inverse of $f^{-1}(x)$ is $f(x)$, it is because $y = (f^{-1})^{-1}(x) \Leftrightarrow x = f^{-1}(y) \Leftrightarrow y = f(x)$.
If we substitute $y = f^{-1}(x)$ or $x = f(y)$ into one another, we would obtain the so called **cancellation identities**:
$f(f^{-1}(x)) = x$, for all x in the domain of $f^{-1}(x)$ or $f^{-1}(f(y)) = y$, for all y in the domain of $f(x)$.
If we interchange the x and y coordinate of a point $P(a, b)$ and denoted the new point as $Q(b, a)$, then each point is the reflection of the other in the line $y = x$. Since the equation $y = f^{-1}(x)$ is equivalent to $x = f(y)$, therefore, the graph of $f^{-1}(x)$ and $f(x)$ are reflections of each other in the line $y = x$.
Fig. 2.8 demonstrates the graph of a function $f(x)$ and $f^{-1}(x)$.
If $f(x)$ and $f^{-1}(x)$ are the same function, then $f(x)$ is called self-inverse function.
$f(x) = x$ and $g(x) = \dfrac{1}{x}$ are self-inverse functions, because their inverse functions equal to themselves.

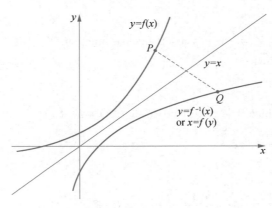

Fig. 2.8

👑 Properties of Inverse Function 反函数的性质

(1) The domain of $f(x)$ is the range of $f^{-1}(x)$.
(2) The range of $f(x)$ is the domain of $f^{-1}(x)$.
(3) The graph of $f^{-1}(x)$ is the reflection of the graph of $f(x)$ in the line $y = x$.
(1) $f(x)$ 的定义域是 $f^{-1}(x)$ 的值域.
(2) $f(x)$ 的值域是 $f^{-1}(x)$ 的定义域.
(3) $f^{-1}(x)$ 的图像是 $f(x)$ 的图像在直线 $y = x$ 上的反射.

Note 注意

$(fg)^{-1} = g^{-1}f^{-1}$; we omit the proof, because it is beyond the scope of this course.

Example 例 2.17

$f(x) = x^2 - 4x + 1$ for $k \leq x \leq 8$.
(a) Find the smallest value for k so the function has an inverse.
(b) For this value of k, find $f^{-1}(x)$, and state the domain and range of $f^{-1}(x)$.

Solution 解

(a) We know a parabola is symmetry about the vertical line $x = \alpha$, where α is the x-coordinate of the vertex.
$f(x) = x^2 - 4x + 1 = (x - 2)^2 - 3 \Rightarrow (2, -3)$ is the vertex.
Therefore, the function $f(x) = x^2 - 4x + 1$ is one-to-one if $x \in [2, +\infty)$ or $x \in (-\infty, 2]$.
In this case, $k = 2$.
(b) $y = (x - 2)^2 - 3$.
$f^{-1}(x): x = (y - 2)^2 - 3 \Rightarrow y = 2 \pm \sqrt{x + 3}$, since the domain for $f(x)$ is $2 \leq x \leq 8$ and the range is $-3 \leq y \leq 33$. Therefore, the domain for $f^{-1}(x)$ is $-3 \leq x \leq 33$ and the range is $2 \leq y \leq 8$.
So the inverse function $f^{-1}(x) = 2 + \sqrt{x + 3}$.

Example 例 2.18

Find the value of k such that $f(x) = \dfrac{x - 6}{x + k}$ is a self-inverse function.

Solution 解

Use the cancellation identity, since it is self-inverse; so $f(x) = f^{-1}(x)$.
$ff(x) = x \Rightarrow ff(0) = 0 \Rightarrow f\left(\dfrac{-6}{k}\right) = 0$.

$f\left(\dfrac{-6}{k}\right) = \dfrac{\dfrac{-6}{k} - 6}{\dfrac{-6}{k} + k} = \dfrac{-6 - 6k}{k^2 - 6} = \dfrac{-6(1 + k)}{(k - \sqrt{6})(k + \sqrt{6})} = 0 \Rightarrow k = -1$.

2.7 Transformation of Function　函数变形

We can apply transformations to the graph of a function to obtain a new graph of a new function which is related to the original function. The simple transformations are translation, stretching and reflecting.
可以对一个函数的图形进行变形，从而得到一个与原函数相关的新函数的图形.简单的变形包括平移、拉伸和翻折.

Horizontal/Vertical Translation　水平/垂直平移

The graph of a function $f(x)$ can be shifted c units horizontally by replacing x with $x - c$ or vertically by replacing y with $y - c$.
To shift a graph c units to the right, replacing x by $x - c$, if $c < 0$, the shift will be to the left.
To shift a graph c units upwards, replacing y by $y - c$, if $c < 0$, the shift will be downward.
将 x 替换为 $x-c$ 时，函数的图形水平平移 c 单位，将 y 替换为 $y-c$ 时，函数的图形垂直平移 c 单位.
如果要将图形向右平移 c 单元，可以将 x 替换为 $x-c$，当 $c<0$ 时，则向左平移.
如果要将图形向上平移 c 单元，可以将 y 替换为 $y-c$，当 $c<0$ 时，则向下平移.

> **Vertical and Horizontal Translation　水平和垂直平移**
> Let $c > 0$ (Fig. 2.9):
> $y = f(x) + c$, shift the graph of $y = f(x)$ at a distance c units upward.
> $y = f(x) - c$, shift the graph of $y = f(x)$ at a distance c units downward.
> $y = f(x + c)$, shift the graph of $y = f(x)$ at a distance c units to the right.
> $y = f(x - c)$, shift the graph of $y = f(x)$ at a distance c units to the left.

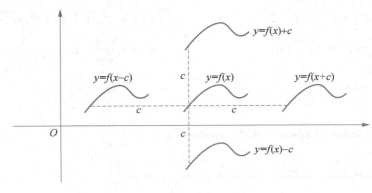

Fig. 2.9

We can use two-dimensional vector to describe the horizontal and vertical translation.

Suppose c is positive, the graph of $y = f(x) + c$ is a translation of the graph of $f(x)$ by the vector $\begin{pmatrix} 0 \\ c \end{pmatrix}$.

The graph of $y = f(x - c)$ is a translation of the graph of $f(x)$ by the vector $\begin{pmatrix} c \\ 0 \end{pmatrix}$.

Suppose a and b are both positive, the graph of $y = f(x - a) + b$, shift the graph of $y = f(x)$ b units upward and a units to the right is equivalent to translate the graph of $f(x)$ by vector $\begin{pmatrix} a \\ b \end{pmatrix}$.

Example 2.19

Find the equation of each graph after the given transformation.

(a) $y = 6x^2$ after translation by $\begin{pmatrix} 3 \\ 0 \end{pmatrix}$;

(b) $y = \sqrt{x - 3}$ after translation by $\begin{pmatrix} -2 \\ 0 \end{pmatrix}$;

(c) $y = x^3 - x + 1$ after translation by $\begin{pmatrix} 0 \\ 4 \end{pmatrix}$;

(d) $y = \dfrac{3}{x + 2}$ after translation by $\begin{pmatrix} 0 \\ -1 \end{pmatrix}$;

(e) $y = \dfrac{3}{x + 1} + x^2 - 2x + 1$ after translation by $\begin{pmatrix} 3 \\ 1 \end{pmatrix}$;

(f) $y = \dfrac{4x - 2}{2x + 3}$ after translation by $\begin{pmatrix} -2 \\ 4 \end{pmatrix}$.

Solution

(a) $y = 6(x - 3)^2$.

(b) $y = \sqrt{x + 2 - 3} = \sqrt{x - 1}$.

(c) $y = x^3 - x + 1 + 4 = x^3 - x + 5$.

(d) $y = \dfrac{3}{x + 2} - 1$.

(e) $y = \dfrac{3}{x + 1} + x^2 - 2x + 1 = \dfrac{3}{x + 1} + (x - 1)^2$. After the translation, the equation would be

$$y = \dfrac{3}{(x - 3) + 1} + (x - 3 - 1)^2 = \dfrac{3}{x - 2} + (x - 4)^2.$$

(f) $y = \dfrac{4(x + 2) - 2}{2(x + 2) + 3} + 4 = \dfrac{4x + 6}{2x + 7} + 4 = 2 + 4 - \dfrac{8}{2x + 7} = 6 - \dfrac{8}{2x + 7}$.

Example 2.20

Find the translation of the followings:

(a) $y = x^2 - 3x + 1$ to $y = x^2 - 3x - 2$;

(b) $y = \sqrt{x + 1}$ to $y = \sqrt{x - 4}$;

(c) $y = (x - 2)^2$ to $y = (x + 1)^2 - 6$;

(d) $y = \dfrac{4}{x - 1}$ to $y = \dfrac{4}{x - 5} + 3$.

Solution

(a) It is a vertical translation of three units downwards, so it is $\begin{pmatrix} 0 \\ -3 \end{pmatrix}$.

(b) It is a horizontal translation of five units to the right, so it is $\begin{pmatrix} 5 \\ 0 \end{pmatrix}$.

(c) It is a vertical translation of six units downwards and a horizontal translation of three units to the left, so it is $\begin{pmatrix} -3 \\ -6 \end{pmatrix}$.

(d) It is a vertical translation of three units upwards and a horizontal translation of four units to the right, so it is $\begin{pmatrix} 4 \\ 3 \end{pmatrix}$. ∎

2.8　Stretch/Reflection　拉伸/翻折

Consider the graph of $y = cf(x)$, where c is a constant. If $c > 1$, then it is a vertical stretch by a factor of c from the graph of $f(x)$; similarly, the graph of $y = \dfrac{1}{c} f(x)$ is a vertical compression by a factor of c.

The graph of $y = -f(x)$ is a reflection of the graph of $y = f(x)$ about x-axis.

The graph of $y = f(-x)$ is a reflection of the graph of $y = f(x)$ about y-axis.

The graph of $y = f(cx)$, where c is a constant. If $c > 1$, then it is a horizontal compression by a factor of c from the graph of $f(x)$; similarly, the graph of $y = f\left(\dfrac{1}{c} x\right)$ is a horizontal stretch by a factor of c (Fig. 2.10).

现有 $y = cf(x)$ 的图形，其中 c 是常数。如果 $c > 1$，可以把该图形看作 $f(x)$ 的图形垂直拉升 c 倍；类似地，可以把图形 $y = \dfrac{1}{c} f(x)$ 看作 $f(x)$ 的图形垂直压缩为 c 分之一。

$y = -f(x)$ 的图形是 $f(x)$ 的图形关于 x 轴的翻折。

$y = f(-x)$ 的图形是 $f(x)$ 的图形关于 y 轴的翻折。

现有 $y = f(cx)$ 的图形，其中 c 是常数。如果 $c > 1$，可以把该图形看作 $f(x)$ 的图形水平压缩为 c 分之一；类似地，可以把图形 $y = f\left(\dfrac{1}{c}\right) x$ 看作 $f(x)$ 的图形水平拉升 c 倍（图 2.10）。

Note 注意

Vertical compression by a factor of c is equivalent to a vertical stretch by a factor of $\dfrac{1}{c}$.

Horizontal compression by a factor of c is equivalent to a horizontal stretch by a factor of $\dfrac{1}{c}$.

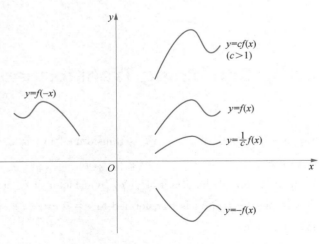

Fig. 2.10

2.9 Special Reflection in Straight Lines 直线上的特殊翻折

When we substitute $-x$ in place of x in a function $y = f(x)$, it corresponds to a reflection about y-axis from the graph of $f(x)$.

When we substitute $-y$ in place of y in a function $y = f(x)$, it corresponds to a reflection about x-axis from the graph of $f(x)$.

When we substitute $k - x$ in place of x in a function $y = f(x)$, it corresponds to a reflection about the line $x = \dfrac{k}{2}$ from the graph of $f(x)$.

When we substitute $k - y$ in place of y in a function $y = f(x)$, it corresponds to a reflection about the line $y = \dfrac{k}{2}$ from the graph of $f(x)$.

Interchange x and y [The inverse Function of $f(x)$] corresponds to a reflection about the line $y = x$ from the graph of $f(x)$.

当用 $-x$ 替换 $y = f(x)$ 函数的 x 时，则形成该函数关于 y 轴的翻折.

当用 $-y$ 替换 $y = f(x)$ 函数的 y 时，则形成该函数关于 x 轴的翻折.

当用 $k - x$ 替换 $y = f(x)$ 函数的 x 时，则形成该函数关于直线 $x = \dfrac{k}{2}$ 的翻折.

当用 $k - y$ 替换 $y = f(x)$ 函数的 y 时，则形成该函数关于直线 $y = \dfrac{k}{2}$ 的翻折.

当将 x 和 y 相互替换时 [$f(x)$ 的反函数]，则形成该函数关于直线 $y = x$ 的翻折.

2.10 Combining Transformation 组合变形

Suppose the graph of $y = f(x)$ is transformed to $y = af(x) + b$, where a and b are positive constants. The steps are：

Vertical stretch of the graph of $f(x)$ by a factor of a, and then a vertical translation by b units upwards.

The graph of $y = f(x)$ is transformed to $y = f(cx + d)$, where c and d are positive constants.

The steps are：

Rewrite $y = f\left[c\left(x + \dfrac{d}{c} \right) \right]$, since c and d are positive；so it means we have a horizontal translation to the left by $\dfrac{d}{c}$ units, and then a horizontal stretch by a factor of $\dfrac{1}{c}$.

Note 注意

For vertical stretch and vertical translation, it follows the usual order of the operations of the equation. While, horizontal stretch and horizontal translation, it follows the reversing order of the operations of the equation. It means horizontal translation first then horizontal stretch.

Example 例 ⟨2.21⟩

Given that $f(x) = x^2 - 2x + 3$, find the following functions after transformations.

(a) Translation $\begin{bmatrix} 0 \\ -1 \end{bmatrix}$, followed by a vertical stretch by a factor of 4.

(b) Vertical stretch by a factor of 4, followed by a translation $\begin{bmatrix} 0 \\ -1 \end{bmatrix}$.

(c) Reflection about x-axis followed by a translation $\begin{bmatrix} 0 \\ 5 \end{bmatrix}$.

(d) Reflection about x-axis followed by a translation $\begin{bmatrix} 0 \\ 5 \end{bmatrix}$ followed by a vertical stretch by a factor of 4.

Solution 解

(a) $y = 4[f(x) - 1] = 4(x^2 - 2x + 3 - 1) = 4x^2 - 8x + 8$.
(b) $y = 4f(x) - 1 = 4(x^2 - 2x + 3) - 1 = 4x^2 - 8x + 11$.
(c) $y = -f(x) + 5 = -(x^2 - 2x + 3) + 5 = -x^2 + 2x + 2$.
(d) $y = 4[-f(x) + 5] = 4[-x^2 + 2x + 2] = -4x^2 + 8x + 8$.

Example 例 ⟨2.22⟩

Consider a function $f(x) = ax + b$, where a and b are constants. The graph of the function is transformed by a sequence of transformation as follows:

Translation by $\begin{bmatrix} 1 \\ 2 \end{bmatrix}$ then a reflection about x-axis; followed by a horizontal stretch with factor of $\frac{1}{3}$.

The resulting function is $g(x) = 4 + 15bx$. Find the values of a and b.

Solution 解

Translation by $\begin{bmatrix} 1 \\ 2 \end{bmatrix}$: $a(x-1) + b + 2$. Reflection about x-axis: $-[a(x-1) + b + 2] = -ax + a - b - 2$.

Horizontal stretch with factor of $\frac{1}{3}$: $-a(3x) + a - b - 2 = 4 + 15bx \Rightarrow -3a = 15b$, $a - b - 2 = 4 \Rightarrow b = -1$, $a = 5$.

Example 例 ⟨2.23⟩

Consider a function $f(x) = ax^2 + bx + c$, where a, b and c are constants. The graph of the function is transformed by a sequence of transformation as follows:

Reflection about y-axis and then translation by $\begin{bmatrix} -1 \\ 3 \end{bmatrix}$ followed by a horizontal stretch with factor of 2. The resulting function is $g(x) = 4x^2 + 16x - 6$. Find the values of a, b and c.

Solution 解

Reflection about y-axis: $a(-x)^2 + b(-x) + c = ax^2 - bx + c$.

Translation by $\begin{bmatrix} -1 \\ 3 \end{bmatrix}$: $a(x+1)^2 - b(x+1) + c + 3$.

Horizontal stretch by a factor of 2: $a\left(\dfrac{x}{2} + 1\right)^2 - b\left(\dfrac{x}{2} + 1\right) + c + 3$

$\Rightarrow a\left(\dfrac{x}{2} + 1\right)^2 - b\left(\dfrac{x}{2} + 1\right) + c + 3 = 4x^2 + 16x - 6 \Rightarrow a\dfrac{x^2}{4} + ax + a - b\dfrac{x}{2} - b + c + 3 = 4x^2 + 16x - 6$

$\Rightarrow \dfrac{a}{4} = 4,\ a - \dfrac{b}{2} = 16,\ a - b + c + 3 = -6 \Rightarrow a = 16,\ b = 0,\ c = -25.$ ∎

Example 例 2.24

Given that $f(x) = 4 - x$, and we define $f^2 = ff$, $f^3 = fff$ and so on (the composite function). Find the value of

(a) $f^2(5)$; (b) $f^3(5)$;

(c) $f^{13}(5)$; (d) $f^n(5)$, where n is a positive integer.

Solution 解

(a) $ff(x) = 4 - (4 - x) = x \Rightarrow ff(5) = 5$.

(b) $f^3(x) = ff^2(x) = f(x) = 4 - x \Rightarrow f^3(5) = 4 - 5 = -1$.

(c) $f^{13}(5) = f^{11}(5) = f^9(5) = \cdots = f^3(5) = f(5) = -1$.

(d) $f^n(5) = \begin{cases} -1, & \text{when } n \text{ is odd}, \\ 5, & \text{when } n \text{ is even}. \end{cases}$ ∎

Example 例 2.25

Functions f and g are defined for $x \in \mathbf{R}$ by
$$f: x \mapsto 4x - 3,$$
$$g: x \mapsto 3x - 1 - x^2.$$

(a) Find the coordinates of the points of intersection of $y = f(x)$ and $y = g(x)$.

(b) Find an expression for $fg(x)$ and deduce the range of fg.

The function h is defined by $h: x \mapsto 3x - 1 - x^2$ for $x \geqslant c$.

(c) Find the smallest value of c for which h has an inverse.

Solution 解

(a) $4x - 3 = 3x - 1 - x^2 \Rightarrow x^2 + x - 2 = 0 \Rightarrow (x-1)(x+2) = 0 \Rightarrow x = 1, -2$.

When $x = 1 \Rightarrow y = 4 - 3 = 1 \Rightarrow$ Coordinates: $(1, 1)$.

When $x = -2 \Rightarrow y = 4 \times (-2) - 3 = -11 \Rightarrow$ Coordinates: $(-2, -11)$.

(b) $fg(x) = 4(3x - 1 - x^2) - 3 = 12x - 4 - 4x^2 - 3 = -4x^2 + 2x - 7$

$= -4\left(x^2 - 3x + \dfrac{9}{4} - \dfrac{9}{4}\right) - 7 = -4\left(x - \dfrac{3}{2}\right)^2 + 2$; so the range is $y \leqslant 2$.

(c) $h(x) = 3x - 1 - x^2 = -(x^2 - 3x) - 1 = -\left(x - \dfrac{3}{2}\right)^2 + \dfrac{5}{4}$.

This function has inverse when $x \geqslant \dfrac{3}{2}$ or $x \leqslant \dfrac{3}{2}$, so the smallest value for c is $\dfrac{3}{2}$. ∎

Example 例 ⟨2.26⟩

The function f is defined by $f: x \to -3x^2 - 2x + 4$ for $x \in \mathbf{R}$.

(a) Express $-3x^2 - 12x + 4$ in the form $a(x + b)^2 + c$, where a, b and c are constant.

(b) Find the coordinates of the vertex on the curve $y = f(x)$.

The function h is defined by $h: x \to -3x^2 - 12x + 4$ for $x \leqslant c$.

(c) State the smallest value of c for which h has an inverse. For this value of c, find $h^{-1}(x)$.

Solution 解

(a) $-3x^2 - 12x + 4 = -3(x^2 + 4x) + 4 = -3(x^2 + 4x + 4 - 4) + 4 = -3(x + 2)^2 + 12 + 4$

$= -3(x + 2)^2 + 16$.

(b) $(-2, 16)$.

(c) The smallest value of $c = -2$.

(d) $h^{-1}: x = -3y^2 - 12y + 4 = -3(y + 2)^2 + 16 \Rightarrow (y + 2)^2 = \dfrac{x - 16}{-3}$.

Because the range of h^{-1} is the domain of h, so we need $h^{-1} \leqslant -2$.

Therefore $h^{-1}(x) = -\sqrt{\dfrac{x - 16}{-3}} - 2$. ∎

Example 例 ⟨2.27⟩

The one-one function f is defined by $f(x) = x^2 - 6x + 12$ for $x \geqslant k$, where k is a constant.

(a) Express $x^2 - 6x + 12$ in the form $(x + a)^2 + b$, where a and b are constant.

(b) State the smallest possible value of k.

(c) Solve the equation $ff(x) = 84$.

(d) When $k = 6$, find an expression for $f^{-1}(x)$ and state the domain of f^{-1}.

Solution 解

(a) $x^2 - 6x + 12 = x^2 - 6x + 9 - 9 + 12 = (x - 3)^2 + 3$.

(b) The smallest value of $k = 3$.

(c) $ff(x) = [(x - 3)^2 + 3 - 3]^2 + 3 = (x - 3)^4 + 3 = 84 \Rightarrow (x - 3)^4 = 81 \Rightarrow x - 3 = \pm 3$

$\Rightarrow x = 6$ only because $x \geqslant 3$.

(d) $f^{-1}: x = y^2 - 6y + 12 = (y - 3)^2 + 3 \Rightarrow (y - 3)^2 = 3 - x$.

Because the range of f^{-1} is the domain of f, so we need $f^{-1} \geq 6$.

Therefore $f^{-1}(x) = \sqrt{3-x} + 3$.

$x \geq 6 \Rightarrow x - 3 \geq 3 \Rightarrow (x-3)^2 \geq 9 \Rightarrow (x-3)^2 + 3 \geq 12$.

The domain of f^{-1} is the range of f, so the domain of $f^{-1} \geq 12$.

Example 例 ⟨2.28⟩

Functions f and g are defined by

$f(x) = \dfrac{12}{x-4} + 4$ for $x > 4$,

$g(x) = \dfrac{12}{x-4} + 4$ for $4 < x < 6$.

(a) State the range of the function f.
(b) State the range of the function g.
(c) State the range of the function fg.

Solution 解

(a) $x > 4 \Rightarrow x - 4 > 0 \Rightarrow \dfrac{1}{x-4} > 0 \Rightarrow \dfrac{12}{x-4} > 0 \Rightarrow \dfrac{12}{x-4} + 4 > 4$.

Hence, the range of the function $f > 4$.

(b) $4 < x < 6 \Rightarrow 0 < x - 4 < 2 \Rightarrow \dfrac{1}{x-4} > \dfrac{1}{2} \Rightarrow \dfrac{12}{x-4} > 6 \Rightarrow \dfrac{12}{x-4} + 4 > 10$.

Hence, the range of the function $g > 10$.

(c) $fg(x) = \dfrac{12}{\left(\dfrac{12}{x-4} + 4\right) - 4} + 4 = \dfrac{12}{\dfrac{12}{x-4}} + 4 = x - 4 + 4 = x$.

So the range for $fg(x)$ is $4 < x < 6$.

Example 例 ⟨2.29⟩

Functions f and g are defined for $x > 1$ by

$$f: x \to \dfrac{1}{x^2 - 1}, \ g: x \to 4x - 5.$$

(a) Find an expression for $f^{-1}(x)$ and state the domain of f^{-1}.
(b) Solve the equation $gg(x) = 7$.
(c) Solve the equation $fg(x) = \dfrac{1}{8}$.

Solution 解

(a) $f^{-1}: x = \dfrac{1}{y^2 - 1} \Rightarrow y^2 - 1 = \dfrac{1}{x} \Rightarrow y^2 = \dfrac{1}{x} + 1 \Rightarrow y = \sqrt{\dfrac{1}{x} + 1}$, domain is $x > 0$.

(b) $gg(x) = 4(4x - 5) - 5 = 7 \Rightarrow 16x - 20 = 12 \Rightarrow 16x = 32 \Rightarrow x = 2$.

(c) $fg(x) = \dfrac{1}{(4x-5)^2 - 1} = \dfrac{1}{8} \Rightarrow (4x-5)^2 - 1 = 8 \Rightarrow (4x-5)^2 = 9 \Rightarrow 4x - 5 = \pm 3$

$\Rightarrow x = 2$ only, because the domain $x > 1$. ∎

Example 例 ⟨2.30⟩

A function f is defined by $f: x \to 3x + 6$ for $x \in \mathbf{R}$.

(a) Find an expression for $f^{-1}(x)$ and solve the equation $f(x) = f^{-1}(x)$.

(b) Sketch, on the same diagram, the graphs of $y = f(x)$ and $y = f^{-1}(x)$, making clear the relationship between the graphs.

Solution 解

(a) $f^{-1}: x = 3y + 6 \Rightarrow 3y = x - 6 \Rightarrow y = \dfrac{x-6}{3}$.

$3x + 6 = \dfrac{x-6}{3} \Rightarrow 9x + 18 = x - 6 \Rightarrow x = -3$.

(b) $f(x)$ and $f^{-1}(x)$ are shown in Fig. 2.11. ∎

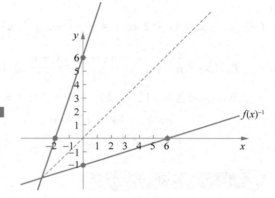

Fig. 2.11

Example 例 ⟨2.31⟩

The functions f and g are defined by

$$f(x) = \dfrac{3}{x^2 - 9} \text{ for } x < -3,$$

$$g(x) = x^2 + 4 \text{ for } x > 0.$$

(a) Find an expression for $f^{-1}(x)$ and $g^{-1}(x)$.

(b) Solve the equation $gf(x) = 13$.

Solution 解

(a) $f^{-1}: x = \dfrac{3}{y^2 - 9} \Rightarrow y^2 - 9 = \dfrac{3}{x} \Rightarrow y^2 = 9 + \dfrac{3}{x} \Rightarrow y = -\sqrt{9 + \dfrac{3}{x}}$.

Because the range of f^{-1} is the domain of f, so we need $f^{-1} < -3$, therefore $f^{-1} = -\sqrt{9 + \dfrac{3}{x}}$.

$g^{-1}: x = y^2 + 4 \Rightarrow y^2 = 4 - x \Rightarrow y = \sqrt{4-x}$.

Because the range of f^{-1} is the domain of f, so we need $f^{-1} > 0$, therefore $f^{-1} = \sqrt{4-x}$.

(b) $gf(x) = \left(\dfrac{3}{x^2-9}\right)^2 + 4 = 13 \Rightarrow \left(\dfrac{3}{x^2-9}\right)^2 = 9 \Rightarrow \dfrac{3}{x^2-9} = \pm 3 \Rightarrow x^2 - 9 = \pm 1$

$\Rightarrow x^2 = 10$ or $8 \Rightarrow x = \pm\sqrt{10}$ or $\pm 2\sqrt{2}$.

Since $x < -3 \Rightarrow x = -\sqrt{10}$ only. ∎

Example 例 ⟨2.32⟩

The function f is defined by $f: x \to \dfrac{1}{4-3x}$ for $x \in \mathbf{R}, x \neq \dfrac{4}{3}$.

(a) Find an expression for $f^{-1}(x)$.

The function g is defined by $g: x \to 3x + k$ for $x \in \mathbf{R}$, where k is a constant.

(b) Find the value of k for which $gf(1) = 5$.

(c) Find the set of values of k for the equation $f^{-1}(x) = g^{-1}(x)$ has no real roots.

Solution 解

(a) $f^{-1}: x = \dfrac{1}{4-3y} \Rightarrow 4 - 3y = \dfrac{1}{x} \Rightarrow y = \dfrac{4x-1}{3x} \Rightarrow f^{-1}(x) = \dfrac{4x-1}{3x}$.

(b) $gf(1) = 3 \times \dfrac{1}{4-3\times 1} + k = 5 \Rightarrow k = 2.$

(c) $g^{-1}: x = 3y + 2 \Rightarrow y = \dfrac{x-2}{3} \Rightarrow g^{-1}(x) = \dfrac{x-2}{3}$.

$f^{-1}(x) = g^{-1}(x) \Rightarrow \dfrac{4x-1}{3x} = \dfrac{x-k}{3} \Rightarrow 12x - 3 = 3x^2 - 3kx \Rightarrow 3x^2 + (12-3k)x + 3 = 0.$

We need $\Delta = (12-3k)^2 - 4 \times 3 \times 3 = 144 - 72k + 9k^2 - 36 = 9k^2 - 72k + 108 < 0$

$\Rightarrow k^2 - 8k + 12 < 0$

$\Rightarrow 2 < k < 6.$

Example 例 ⟨2.33⟩

(a) Express $4x^2 - 12x + 15$ in the form $(ax+b)^2 + c$, where a, b and c are constants.

The function f is defined by $f(x) = 4x^2 - 12x + 15$ for $x \geq k$, where k is a constant.

(b) State the smallest value of k for which f is a one-one function.

(c) For this value of k, obtain an expression for $f^{-1}(x)$, and state the domain of $f^{-1}(x)$.

Solution 解

(a) $4x^2 - 12x + 15 = (2x-3)^2 - 9 + 15 = (2x-3)^2 + 6.$

(b) The smallest value of $k = \dfrac{3}{2}$.

(c) $f^{-1}: x = (2y-3)^2 + 6 \Rightarrow (2y-3)^2 = x - 6 \Rightarrow 2y - 3 = \sqrt{x-6}$

$\Rightarrow y = \dfrac{\sqrt{x-6}+3}{2}$. Because the range of f^{-1} is the domain of f,

so we need $f^{-1} \geq \dfrac{3}{2}$; therefore $f^{-1} = \dfrac{\sqrt{x-6}+3}{2}$, the domain of f^{-1} is $x \geq 6$.

Example 例 ⟨2.34⟩

The functions f and g are defined for $x \geq 0$ by

$$f: x \to 4x^2 + 5, \quad g: x \to 2x + 1.$$

(a) Show that $gf(x) = 8x^2 + 11$ and obtain an expression for $fg(x)$.

(b) Find an expression $(fg)^{-1}(x)$ and determine the domain of $(fg)^{-1}$.

Solution

(a) $gf(x) = 2(4x^2 + 5) + 1 = 8x^2 + 11$,

$fg(x) = 4(2x + 1)^2 + 5$.

(b) $(fg)^{-1}: x = 4(2y + 1)^2 + 5 \Rightarrow (2y + 1)^2 = \dfrac{x - 5}{4} \Rightarrow y = \dfrac{\sqrt{x - 5} - 2}{4}$.

Since $x \geq 0 \Rightarrow g(x) \geq 1 \Rightarrow fg(x) \geq 9$.

So the domain of $(fg)^{-1}$ is $x \geq 9$.

Example 2.35

The functions f and g are defined by

$$f(x) = \frac{5}{x} \text{ for } x > 0, \quad g(x) = \frac{5}{2x + 1} \text{ for } x \geq 0.$$

(a) Find and simplify an expression for $fg(x)$ and state the range of fg.

(b) Find an expression for $g^{-1}(x)$ and the domain of g^{-1}.

Solution

(a) $fg(x) = \dfrac{5}{\dfrac{5}{2x+1}} = 2x + 1$, the range of fg is $y \geq 1$.

(b) $g^{-1}: x = \dfrac{5}{2y + 1} \Rightarrow 2y + 1 = \dfrac{5}{x} \Rightarrow y = \dfrac{\dfrac{5}{x} - 1}{2} = \dfrac{5 - x}{2x} = \dfrac{5}{2x} - \dfrac{1}{2}$.

Since the domain for $g(x)$ is $x \geq 0 \Rightarrow$ the range for $g^{-1}(x) \geq 0 \Rightarrow \dfrac{5}{2x} - \dfrac{1}{2} \geq 0 \Rightarrow \dfrac{5}{2x} \geq \dfrac{1}{2} \Rightarrow \dfrac{2x}{5} \leq 2 \Rightarrow$

$x \leq 5$. The range for $g(x)$ is $y > 0$, so the domain for $g^{-1}(x)$ is $x > 0$. Hence, the domain for $g^{-1}(x)$ is $0 < x \leq 5$.

Example 2.36

(a) Express $9x^2 + 24x + 20$ in the form $(ax + b)^2 + c$, where a, b and c are constants.

(b) Functions f and g are both defined for $x > 0$. It is given that $f(x) = x^2 + 4$ and $fg(x) = 9x^2 + 24x + 20$. Find $g(x)$.

(c) Find $(fg)^{-1}(x)$ and give the domain of $(fg)^{-1}$.

Solution

(a) $9x^2 + 24x + 20 = (3x + 4)^2 - 16 + 20 = (3x + 4)^2 + 4$.

(b) $fg(x) = (g(x))^2 + 4 = 9x^2 + 24x + 20 = (3x + 4)^2 + 4 \Rightarrow g(x) = 3x + 4$.

(c) $(fg)^{-1}: x = (3y + 4)^2 + 4 \Rightarrow (3y + 4)^2 = x - 4 \Rightarrow 3y + 4 = \sqrt{x - 4} \Rightarrow y = \dfrac{\sqrt{x - 4} - 4}{3}$.

Since the range for fg is $fg > 20$, so the domain for $(fg)^{-1}$ would be $x > 20$.

Example 2.37

The function f is defined by $f: x \to 3x - x^2 + 4$ for $x \in \mathbf{R}$.

(a) Find the set of values of x for which $f(x) \geqslant 0$.

(b) Given that the line $y = mx + 8$ is a tangent to the curve $y = f(x)$, find m.

The function g is defined by $g: x \to 3x - x^2 + 4$ for $x \geqslant k$, where k is a constant.

(c) Express $3x - x^2 + 4$ in the form $a(x+b)^2 + c$, where a, b and c are constants.

(d) State the smallest value of k for which g has an inverse.

(e) For this value of k, find an expression for $g^{-1}(x)$ and find the domain of g^{-1}.

Solution

(a) $f(x) = 3x - x^2 + 4 \geqslant 0 \Rightarrow x^2 - 3x - 4 \leqslant 0 \Rightarrow (x-4)(x+1) \leqslant 0 \Rightarrow -1 \leqslant x \leqslant 4$.

(b) $3x - x^2 + 4 = mx + 8 \Rightarrow x^2 + (m-3)x + 4 = 0$.

We need $D = (m-3)^2 - 4 \times 4 = 0 \Rightarrow (m-3)^2 = 16 \Rightarrow m - 3 = \pm 4 \Rightarrow m = 7$ or $m = -1$.

(c) $3x - x^2 + 4 = -(x^2 - 3x) + 4 = -\left(x - \dfrac{3}{2}\right)^2 + \dfrac{25}{4}$.

(d) The smallest value of k is $\dfrac{3}{2}$.

(e) $g^{-1}: x = 3y - y^2 + 4 = -\left(y - \dfrac{3}{2}\right)^2 + \dfrac{25}{4} \Rightarrow \left(y - \dfrac{3}{2}\right)^2 = \dfrac{25}{4} - x \Rightarrow y = \sqrt{\dfrac{25}{4} - x} + \dfrac{3}{2}$, the domain of g^{-1} is $x \leqslant \dfrac{25}{4}$.

Example 2.38

The function f is such that $f(x) = 4x + 3$ for $x \geqslant 0$. The function g is such that $g(x) = ax^2 - 8$ for $x \leqslant k$, where a and k are constants. The function fg is such that $fg(x) = 8x^2 - 29$ for $x \leqslant k$.

(a) Find the values of a.

(b) Find the greatest possible value of k.

(c) For the greatest possible value of k, find an expression for $f^{-1}(x)$ and state the domain of f^{-1}.

(d) Given that $k = -5$, find the range of fg.

Solution

(a) $fg(x) = 4(ax^2 - 8) + 3 = 4ax^2 - 29 = 8x^2 - 29 \Rightarrow a = 2$.

(b) $g(x) = 2x^2 - 8 \geqslant 0 \Rightarrow x^2 \geqslant 4 \Rightarrow x \geqslant 2$ or $x \leqslant -2$.

So, the gratest possible value of $k = -2$.

(c) $f^{-1}: x = 4y + 3 \Rightarrow y = \dfrac{x-3}{4}$, the domain of f^{-1} is $x \geqslant 3$.

(d) $fg(x) = 8x^2 - 29$ and $k = -5$, so $fg \geqslant 8 \times (-5)^2 - 29 = 171$.

Summary of Key Theories 核心定义总结

(1) A **function** f from a set X to a set Y is a rule that assigns a unique element $f(x)$ in Y (outputs) to a unique element x in X (input), with the property that each input is related to only one output. An alternative name for a function is a **mapping**.

(2) The notation $f: X \to Y$ means f is a function from X to Y, where X is called the **domain** of f, and the set of all values of f taken together is called the **range** of f.

(3) Consider a function $y = f(x)$, then the **independent variable** is x, and y is the **dependent variable**.

(4) A function $f(x)$ is called an even function if it satisfies $f(-x) = f(x)$ $\forall x$ in the domain of $f(x)$ ($\forall$ is the mathematical symbol means "for all"). The graph of an even function is symmetry about the y-axis.

(5) A function $f(x)$ is called an odd function is it satisfies $f(-x) = -f(x)$ $\forall x$ in the domain of $f(x)$. The graph of an odd function is symmetry about the origin.

(6) A function $f(x)$ is called a **one-to-one function** or **injective mapping** if, and only if, for all elements x_1 and x_2 in the domain: if $f(x_1) = f(x_2)$, then $x_1 = x_2$. Or, equivalently, if $x_1 \neq x_2$, then $f(x_1) \neq f(x_2)$.

(7) Let $f: X \text{ a } Y'$ and $g: Y \text{ a } Z$ be functions with the property that the range of f is a subset of the domain of g. We define a new function $g \circ f: X \to Z$ as $g \circ f(x) = g(f(x))$ $\forall x \in X$, where $g(f(x))$ is read "g of f of x". The function $g \circ f(x) = g(f(x))$ is called the composition of f and g.

(8) If $f(x)$ is one-to-one, then it has as inverse function $f^{-1}(x)$. The value of $f^{-1}(x)$ is the unique number y in the domain of $f(x)$ such that $f(y) = x$. It means $y = f^{-1}(x)$ if and only if $x = f(y)$.

(9) If $f(x)$ is one-to-one, then it has as inverse function $f^{-1}(x)$. The value of $f^{-1}(x)$ is the unique number y in the domain of $f(x)$ such that $f(y) = x$. It means $y = f^{-1}(x)$ if and only if $x = f(y)$.

① The domain of $f(x)$ is the range of $f^{-1}(x)$;

② The range of $f(x)$ is the domain of $f^{-1}(x)$;

③ The graph of $f^{-1}(x)$ is the reflection of the graph of $f(x)$ in the line $y = x$.

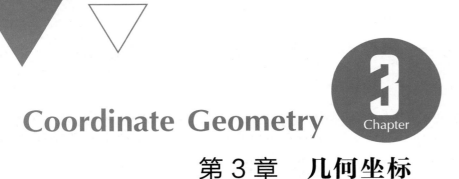

Coordinate Geometry

第 3 章 几何坐标

3.1 Line and the Gradient 直线与斜率

In geometry, a straight line can be constructed by connecting two points $A(x_1, y_1)$ and $B(x_2, y_2)$.

The gradient of a line (also known as slope) is a number that describes the direction and the steepness of the line. The gradient is calculated by finding the ratio of the vertical distance to the horizontal distance between the two points which were connected by the line. The ratio of the vertical distance and the horizontal distance can be expressed as a quotient, and we sometimes called it "rise over run".

在几何中，可以通过连接 $A(x_1, y_1)$ 和 $B(x_2, y_2)$ 两个点获得一条直线. 直线的倾斜度（也称为斜率）是一个用来描述直线的方向和梯度的数字. 梯度通过求直线所连接的两点之间的垂直距离与水平距离之比来计算. 垂直距离和水平距离的比值可以用商来表示，有时称为"高度比长度".

Let the gradient be m, the line joining the points A and B would be $m = \dfrac{y_2 - y_1}{x_2 - x_1}$. The equation of the line $\overline{AB}$ can be determined by thinking there is a general point called it $C(x, y)$; and the gradient between C and A is $\dfrac{y - y_1}{x - x_1} = m$ or the gradient between C and B is $\dfrac{y - y_2}{x - x_2} = m$ to yield the equation $y = m(x - x_1) + y_1$ or $y = m(x - x_2) + y_2$. After simplifying, we can show these two equations are identical, so we can write the equation of the line as $y = mx + b$, where m is the gradient of the line and b is the y-intercept.

Definition 定义

The gradient between two points $A(x_1, y_1)$ and $B(x_2, y_2)$ is $m = \dfrac{y_2 - y_1}{x_2 - x_1}$.

Chapter 3　Coordinate Geometry　　第三章　几何坐标

The equation of the line segment $\overline{AB}$ is given by $\dfrac{y - y_1}{x - x_1} = \dfrac{y_2 - y_1}{x_2 - x_1}$ or $\dfrac{y - y_2}{x - x_2} = \dfrac{y_2 - y_1}{x_2 - x_1}$.

The x-intercept of the line is determined when $y = 0$.

The y-intercept of the line is determined when $x = 0$.

$A(x_1, y_1)$ 和 $B(x_2, y_2)$ 两点之间的斜率是 $m = \dfrac{y_2 - y_1}{x_2 - x_1}$.

线段 $\overline{AB}$ 的方程为 $\dfrac{y - y_1}{x - x_1} = \dfrac{y_2 - y_1}{x_2 - x_1}$ 或者 $\dfrac{y - y_2}{x - x_2} = \dfrac{y_2 - y_1}{x_2 - x_1}$.

当 $y = 0$ 时，可计算该直线在 x 轴上的截距；

当 $x = 0$ 时，可计算该直线在 y 轴上的截距.

Since the gradient, m, describes the direction and steepness of the line, so, when $m > 0$, we said the line is increasing, which means when the value of x gets larger, the value of y gets larger.

When $m < 0$, the line is decreasing which means when the value of x gets larger, the value of y gets smaller.

When $m = 0$ means the value of x gets larger, the value of y remains unchanged. So it is a horizontal line with equation $y = k$, where k is any constant.

Example 例 3.1

Find the gradient and the equation of the line connecting $A(1, 3)$ and $B(-2, -4)$.

Solution 解

Gradient $= \dfrac{3 - (-4)}{1 - (-2)} = \dfrac{7}{3}$, the equation of the line is $\dfrac{y - 3}{x - 1} = \dfrac{7}{3} \Rightarrow y = \dfrac{7}{3}x + \dfrac{2}{3}$.

Example 例 3.2

The line with equation $2y + 3x = 5$ intersects the x-axis at point P and y-axis at point Q, find the coordinates of P and Q.

Solution 解

The coordinate of P is determined when $y = 0 \Rightarrow 3x = 5 \Rightarrow x = \dfrac{5}{3} \Rightarrow P\left(\dfrac{5}{3}, 0\right)$.

The coordinate of Q is determined when $x = 0 \Rightarrow 2y = 5 \Rightarrow y = \dfrac{5}{2} \Rightarrow Q\left(0, \dfrac{5}{2}\right)$.

Example 例 3.3

The point $P(k, -3)$ lies on the line through the points $A(-1, 5)$ and $B(3, 7)$. What is the value of k?

Solution 解

Three points A, B and P lie on the same line, so they are collinear. Therefore, we can use the gradient to find the coordinate of P.

$$\frac{5-(-3)}{-1-k} = \frac{7-5}{3-(-1)} \Rightarrow \frac{8}{-(1+k)} = \frac{2}{4} \Rightarrow -(1+k) = 16 \Rightarrow k = -17.$$

3.2 Parallel and Perpendicular Lines 平行线与垂线

Two lines in xy-plane are parallel if and only they have the same gradient and no points in common. For examples:
$y = 3x - 5$ and $y = 3x - 2$ are parallel, because they have the same gradient and no points in common (Fig. 3.1).
$y = 3x - 5$ and $y = 2x + 1$ are not parallel, because they do not have the same gradient and they would intersect at a point called point of intersection. Any two non-parallel lines in xy-plane would intersect at a point.
When two lines intersect at a right angle, we said they are perpendicular to each other. To find the relationship between the gradients of perpendicular lines, we constructed two congruent triangles as shown in Fig. 3.2.

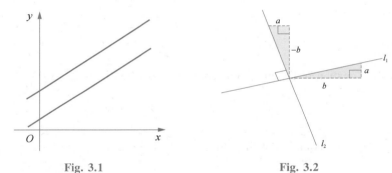

Fig. 3.1   Fig. 3.2

The line l_1 has gradient $\frac{a}{b}$, while the line l_2 has gradient $\frac{-b}{a}$; hence, the product of their gradients equal to -1.

> ♛ **Definition** 定义
>
> Given two lines $l_1: y = m_1 x + b_1$ and $l_2: y = m_2 x + b_2$.
> If $m_1 = m_2$ and $b_1 \neq b_2$, then two lines are parallel.
> If $m_1 \cdot m_2 = -1$ regardless the values of b_i; then two lines are perpendicular.
> 假设有两条线 $l_1: y = m_1 x + b_1$ 和 $l_2: y = m_2 x + b_2$.
> 如果 $m_1 = m_2$ 且 $b_1 \neq b_2$，那么，这两条线相互平行.
> 如果 $m_1 \cdot m_2 = -1$，无论 b_i 为何值，那么，这两条线相互垂直.

Example 例 3.4

If $2x + ay = 3$ and $bx + 4y = 7$ represent the same line. What are the values of a and b?

Solution

Since they represent the same line, so $7(2x + ay) = 7 \cdot 3 = 3(bx + 4y) \Rightarrow 14x + 7ay = 3bx + 12y$. Hence, $14 = 3b \Rightarrow b = \dfrac{14}{3}$ and $7a = 12 \Rightarrow a = \dfrac{12}{7}$.

Example 3.5

Two perpendicular lines l_1 and l_2 have an intersection point $P(5, 1)$. If the value of x-intercept of l_1 is twice of the value of x-intercept of l_2; find the possible sum of x-intercepts of l_1 and l_2.

Solution

Let $A(a, 0)$ and $B(2a, 0)$ be the x-intercepts of l_2 and l_1 respectively.

Since two lines are perpendicular, this implies that the product of the gradient of AP and BP is -1.

$$\dfrac{0-1}{a-5} \cdot \dfrac{0-1}{2a-5} = -1 \Rightarrow 2a^2 - 15a + 26 = 0 \Rightarrow a = \dfrac{15 \pm \sqrt{17}}{4}.$$

The sum of x-intercepts is $3a = 3\left(\dfrac{15 + \sqrt{17}}{4}\right)$ or $3\left(\dfrac{15 - \sqrt{17}}{4}\right)$.

Example 3.6

The straight line $5x - 6y - 30 = 0$ is reflected about the line $y = -x$. Find the equation of the image.

Solution

Method 1

Let $A(0, -5)$ and $B(6, 0)$ be points on the line. So the image would be $A'(5, 0)$ and $B'(0, -6)$.

The gradient of $\overline{AB} = \dfrac{-6-0}{0-5} = \dfrac{6}{5}$, so the equation of the image is $\dfrac{y-0}{x-5} = \dfrac{6}{5} \Rightarrow y = \dfrac{6}{5}x - 6$.

Method 2

The image of (x, y) after reflection is $(-y, -x)$. Therefore, the image of reflection is
$5(-y) - 6(-x) - 30 = 0 \Rightarrow 6x - 5y - 30 = 0$.

Example 3.7

The point A is the reflection of the point $(2, 3)$ in the line $y = 3x - 6$, find the coordinate of A.

Solution

The equation of the perpendicular line passing through A is $\dfrac{y-3}{x-2} = \dfrac{-1}{3}$, the intersection point between the line and the perpendicular line is $(2.9, 2.7)$. Therefore, the coordinate of A would be $(3.8, 2.4)$.

3.3 Length of Line Segment and Midpoint
线段的长度与中点

To find the length between two points in Cartesian coordinate system, we can construct a right triangle and use the Pythagoras' Theorem to find the required length as shown in Fig. 3.3.

The length of $AB = d = \sqrt{(x_1 - x_2)^2 + (y_1 - y_2)^2}$.

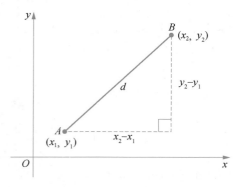

Fig. 3.3

In geometry, the **midpoint** is the middle point of a line segment. It is equidistant from both endpoints, bisects the segment; and it must lie on the line segment.

The midpoint $M(a, b)$ of two points $A(x_1, y_1)$ and $B(x_2, y_2)$ is given by $a = \dfrac{x_1 + x_2}{2}$ and $b = \dfrac{y_1 + y_2}{2}$.

We can verify the result by using the distance formula and the gradient formula. We have to show $\overline{AM} = \overline{BM}$ and their gradients are equal.

The midpoint is equidistant from A and B; it bisects the segment. We now check if $\overline{AM} = \overline{BM}$ is true.

The length of $\overline{AM}$ is $\sqrt{\left(\dfrac{x_1 + x_2}{2} - x_1\right)^2 + \left(\dfrac{y_1 + y_2}{2} - y_1\right)^2} = \sqrt{\left(\dfrac{x_2 - x_1}{2}\right)^2 + \left(\dfrac{y_2 - y_1}{2}\right)^2}$.

The length of $\overline{BM}$ is $\sqrt{\left(\dfrac{x_1 + x_2}{2} - x_2\right)^2 + \left(\dfrac{y_1 + y_2}{2} - y_2\right)^2} = \sqrt{\left(\dfrac{x_2 - x_1}{2}\right)^2 + \left(\dfrac{y_2 - y_1}{2}\right)^2}$.

Therefore, they are equal.

Recall, the gradient of the line segment AB is $\dfrac{\Delta y}{\Delta x} = \dfrac{y_2 - y_1}{x_2 - x_1}$.

Now, using the gradient formula to find the gradient of $\overline{AM}$: $\dfrac{\dfrac{y_1 + y_2}{2} - y_1}{\dfrac{x_1 + x_2}{2} - x_1} = \dfrac{y_2 - y_1}{x_2 - x_1}$,

The gradient of $\overline{BM}$: $\dfrac{y_2 - \dfrac{y_1 + y_2}{2}}{x_2 - \dfrac{x_1 + x_2}{2}} = \dfrac{y_2 - y_1}{x_2 - x_1}$.

Hence, the gradients of $\overline{AM}$ and $\overline{BM}$ are equal.

Chapter 3   Coordinate Geometry   第三章 几何坐标

Theorem 3.1   定理 3.1

The distance between two points $A(x_1, y_1)$ and $B(x_2, y_2)$ is $d = \sqrt{(x_1 - x_2)^2 + (y_1 - y_2)^2}$, and the coordinate of the midpoint $M(a, b)$ is given by $a = \dfrac{x_1 + x_2}{2}$ and $b = \dfrac{y_1 + y_2}{2}$.

点 $A(x_1, y_1)$ 和点 $B(x_2, y_2)$ 之间的距离为 $d = \sqrt{(x_1 - x_2)^2 + (y_1 - y_2)^2}$，并且中点 $M(a, b)$ 的坐标为 $a = \dfrac{x_1 + x_2}{2}$ 和 $b = \dfrac{y_1 + y_2}{2}$。

Proof   证明

It is proved above.

Example 例 3.8

Given two parallel lines $l_1: y = 2x + 5$ and $l_2: y = 2x - 3$, find the distance between these two lines.

Solution 解

Let $A(0, 5)$ be a point on l_1.

We first find a line l_3 which is perpendicular to l_1 and passing through point A.

$l_3: \dfrac{y - 5}{x - 0} = \dfrac{-1}{2} \Rightarrow y = \dfrac{-1}{2}x + 5$.

Then we find the intersection between l_3 and l_2,

$\begin{cases} y = \dfrac{-1}{2}x + 5 \\ y = 2x - 3 \end{cases} \Rightarrow x = \dfrac{16}{5}, y = \dfrac{17}{5}$;

so the intersection point $B\left(\dfrac{16}{5}, \dfrac{17}{5}\right)$. The shortest distance between the line would be the distance of

$\overline{AB} = \sqrt{\left(0 - \dfrac{16}{5}\right)^2 + \left(5 - \dfrac{17}{5}\right)^2} = \dfrac{\sqrt{320}}{5} = \dfrac{8}{\sqrt{5}}$.

Example 例 3.9

Determine the coordinate of the points P which divides the line segment $\overline{AB}$ internally with ratio $5:3$, where $A(1, 3)$ and $B(24, -9)$.

Solution 解

Let $P(a, b)$, then constructing two similar triangles ABC and APQ, where $C(1, -9)$ and $Q(1, b)$.

From the properties of similar triangles, we have $\dfrac{a - 1}{24 - 1} = \dfrac{5}{8} \Rightarrow a = \dfrac{123}{8}$ and $\dfrac{3 - b}{3 - (-9)} = \dfrac{5}{8} \Rightarrow b = \dfrac{-9}{2}$.

Example 例 3.10

Find the equation of the set of points that are equidistant from two points $A(5, 0)$ and $B(0, -3)$.

Solution 解

Method 1 (Using Distance Formula)

Let $C(x, y)$ be the points that are equidistance from $A(5, 0)$ and $B(0, -3)$. Therefore,
$$\sqrt{(x-5)^2 + y^2} = \sqrt{x^2 + (y+3)^2} \Rightarrow x^2 - 10x + 25 + y^2 = x^2 + y^2 + 6y + 9 \Rightarrow 6y + 10x = 16.$$

Method 2 (Using Gradient Property)

The perpendicular bisector of line segment $\overline{AB}$ would be the equation for equidistance from points A and B. The midpoint of $\overline{AB}$ is $M\left(\dfrac{5+0}{2}, \dfrac{0+-3}{2}\right)$, and the gradient of $\overline{AB}$ is $\dfrac{-3-0}{0-5} = \dfrac{3}{5}$. The equation of the perpendicular bisector would have a gradient of $\dfrac{-5}{3}$ passing through M. Hence, the equation is $\dfrac{y - \left(-\dfrac{3}{2}\right)}{x - \dfrac{5}{2}} = \dfrac{-5}{3} \Rightarrow 6y + 10x = 16.$

Example 例 3.11

Given three straight lines l_1, l_2 and l_3; their gradients are $\dfrac{1}{4}$, $\dfrac{1}{5}$ and $\dfrac{1}{6}$, respectively. The sum of their x-intercepts is 45; and they all have the same y-intercept. Find the value of the y-intercept.

Solution 解

Let the equations of l_1, l_2 and l_3 be $y = \dfrac{x}{4} + b$, $y = \dfrac{x}{5} + b$ and $y = \dfrac{x}{6} + b$, respectively. So the x-intercepts for l_1, l_2 and l_3 are $-4b$, $-5b$ and $-6b$, respectively.

Hence, $-4b + -5b + -6b = 45 \Rightarrow -15b = 45 \Rightarrow b = -3$.

Therefore, the y-intercept is -3.

3.4 The Properties of Circle 圆的性质

We should begin this section with some revisions for the circle's theorems from IGCSE Math.

Definition 定义

A tangent to a circle is a line that just touches the circle at one point.

圆的切线就是与圆在一点相切的直线.

Theorem 3.2 定理 3.2

The triangle inscribed in a semi-circle with diameter being one of its sides is a right triangle. That is, △ACB is a right triangle (Fig. 3.4).

与半圆内接、以直径为边的三角形是直角三角形.也就是说，△ACB 是一个直角三角形(图 3.4).

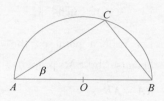

Fig. 3.4

Proof 证明

$\overline{OA} = \overline{OB} = \overline{OC}$ because they are radius of the circle.

So △OAC and △OBC are isosceles triangles.

This implies $\alpha_1 = \alpha_2$ and $\beta_1 = \beta_2$.

The sum of angles in △ABC equal to 180°.

$\alpha_1 + \alpha_2 + \beta_1 + \beta_2 = 180 \Rightarrow 2(\alpha_2 + \beta_2) = 180 \Rightarrow \alpha_2 + \beta_2 = 90$.

So △ACB is a right triangle (Fig.3.5).

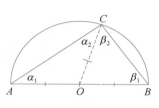

Fig. 3.5

Theorem 3.3 定理 3.3

Three points A, B and C lie on a circle to form a right triangle where $\angle ABC = 90°$. Then $\overline{AC}$ is the diameter of the circle (Fig. 3.6).

若一个圆上的三个点形成一个直角三角形，其中，$\angle ABC = 90°$，则 $\overline{AC}$ 是圆的直径(图 3.6).

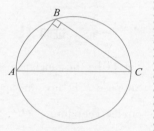

Fig. 3.6

Proof 证明

Suppose $\overline{AD}$ is the diameter of the circle with center M. Then △ABD is a right triangle by Theorem 3.2.

Also, C lies on the circle to make △ABC a right triangle.

$\angle CBD = 90° - 90° = 0° \Rightarrow C$ and D are same point.

Therefore, $\overline{AC}$ is the diameter (Fig. 3.7).

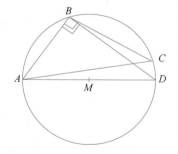

Fig. 3.7

Theorem 3.4 定理 3.4

The perpendicular bisector of a chord of a circle passing through the center of the circle.

圆的弦的垂直平分线通过圆心.

Proof 证明

Let X be a point on the perpendicular bisector of $\overline{AB}$, so $\triangle XAM \cong \triangle XBM$ because the SAS property.

$\overline{XA} = \overline{XB}$.

We can move X along the line $\overline{PM}$ such that $\overline{XA} = \overline{XB}$ = radius of the circle; therefore, X would be the center.

Hence, the perpendicular bisector of the chord would passing through the center of the circle (Fig. 3.8). ■

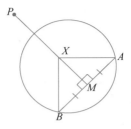

Fig. 3.8

Theorem 3.5 定理 3.5

The tangent to a circle at a point is perpendicular to the radius $\angle OAT = 90°$ at the point of intersection (Fig. 3.9).

圆的切线垂直于交点处的半径，在交点处 $\angle OAT = 90°$（图 3.9）。

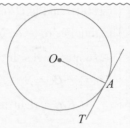

Fig. 3.9

Proof 证明

Let $\overline{AT}$ be the tangent to a circle centered at O.

Suppose P is any other point on the tangent line which is not A and not touch the circle. Therefore, $|\overline{OP}| > |\overline{OA}|$; so $\overline{OA}$ is the shortest distance from O to the tangent line, and $\overline{OA}$ is the radius of the circle.

From fundamental result of geometry, the shortest distance is the perpendicular distance (Fig. 3.10); hence, $\angle OAT = 90°$. ■

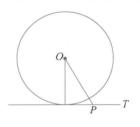

Fig. 3.10

Theorem 3.6 定理 3.6

Two different tangents of a circle from an external point would be equal in length $\overline{AP} = \overline{BP}$, $\angle APO = \angle BPO$, and the line join the point to the center of the circle bisects the angle (Fig. 3.11).

从外部同一点出发的圆的两条不同的切线满足 $\overline{AP} = \overline{BP}$，$\angle APO = \angle BPO$，并且该点和圆心的连线平分这个角（图 3.11）。

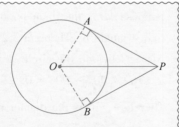

Fig. 3.11

Proof 证明

$\angle OAP = \angle OBP = 90°$ by Theorem 3.5. $\overline{OA} = \overline{OB}$ because they are radius of the circle and $\overline{OP} = \overline{OP}$. Therefore, $\triangle AOP \cong \triangle BOP$.

Hence, $\overline{AP} = \overline{BP}$ and $\angle APO = \angle BPO$. ■

Chapter 3 Coordinate Geometry   第三章 几何坐标

Theorem 3.7   定理 3.7

The angle at the center of a circle is twice the angle on the circle subtended by the same arc (Fig. 3.12). $\angle AOB = 2\angle ACB$.

圆弧所对的圆心角是圆周角的两倍(图 3.12)，$\angle AOB = 2\angle ACB$.

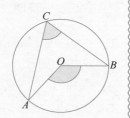

Fig. 3.12

Proof   证明

$\overline{OA} = \overline{OB} = \overline{OC}$ (radius of the circle).

So both $\triangle AOC$ and $\triangle BOC$ are isosceles triangles.

Hence, $\alpha_1 = \alpha_2$ and $\beta_1 = \beta_2$.

Also, $\angle AOX = 2\alpha$ and $\angle BOX = 2\beta$ because they are exterior angles of $\triangle AOC$ and $\triangle BOC$ (Fig. 3.13).

Therefore, $\angle AOB = 2\alpha + 2\beta = 2(\alpha + \beta) = 2\angle ACB$.

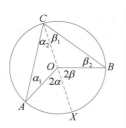

Fig. 3.13

Theorem 3.8   定理 3.8

The angles subtended by the same arc on a circle are equal (Fig. 3.14). $\angle ADB = \angle ACB$.

与同一圆弧相对的圆周角的度数相等(图 3.14)，$\angle ADB = \angle ACB$.

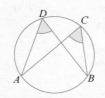

Fig. 3.14

Proof   证明

Let O be the center of the circle (Fig. 3.15), and let $\angle AOB = \alpha$; then by Theorem 3.7, we have $\angle ADB = \dfrac{\alpha}{2}$ and $\angle ACB = \dfrac{\alpha}{2}$.

Hence, $\angle ADB = \angle ACB$.

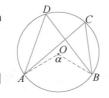

Fig. 3.15

Theorem 3.9   定理 3.9

The angle between a tangent and a chord at the point of intersection is equal to the angle subtended by the chord in the alternate segment (Fig. 3.16). $\angle SAB = \angle BCA$.

在一个点上，切线和弦形成的夹角等于内错弓形的圆周角(图 3.16)，$\angle SAB = \angle BCA$.

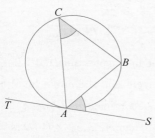

Fig. 3.16

Proof 证明

Let O be the center of the circle, draw $\overline{AOX}$ and $\overline{BX}$.
By Theorem 3.5, $\angle XAS = 90°$.
By Theorem 3.3, $\angle XBA = 90°$.
Let $\angle BAS = \alpha°$, so $\angle BAX = 90° - \alpha°$.
Use $\triangle ABX$, we know $\angle BXA = 90° - (90° - \alpha°) = \alpha°$ and $\angle BXA = \angle BCA$ because they are angles subtended by the same arc (Theorem 3.8) (Fig. 3.17). Hence, $\angle SAB = \angle BCA$.

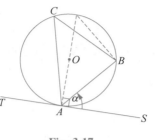

Fig. 3.17

Example 例 3.12

Given a circle with radius r, and it intersects the x-axis at x_1 and 8; where $x_1 < 8$. Furthermore, the y-axis is a tangent line to the circle and their intersection is $(0, 2)$. Determine the value of r (Fig. 3.18).

Solution 解

Let the center of the circle be C (Fig. 3.19). Since y-axis is tangent to the circle, therefore, $\overline{CQ}$ is perpendicular to y-axis. Hence, the coordinate of C is $(r, 2)$. Draw a perpendicular line $\overline{CS}$ and construct a right triangle CSP.

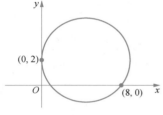 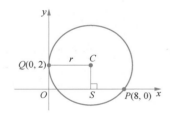

Fig. 3.18　　　　　　　　　Fig. 3.19

Using Pythagoras' Theorem, we have $\overline{CS}^2 + \overline{SP}^2 = \overline{CP}^2 \Rightarrow r^2 = 2^2 + (8-r)^2 \Rightarrow r = \dfrac{17}{4}$.

Example 例 3.13

A circle with center located at $(4, 0)$, and the line $y = x$ is tangent to the circle. Determine the radius of the circle (Fig. 3.20).

Solution 解

Let the center be $D(4, 0)$, and the intersection between the tangent and the circle be G. Since G lies on the line $y = x$, so let the coordinate be $G(a, a)$.
By Theorem 3.5, $\angle OGD$ is $90°$.
Use Pythagoras' Theorem,

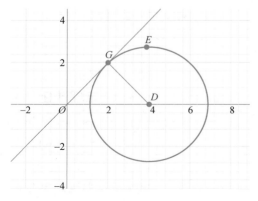

Fig. 3.20

$\overline{OG}^2 + \overline{GD}^2 = \overline{OD}^2 \Rightarrow a^2 + a^2 + (4-a)^2 + a^2 = 16 \Rightarrow 4a(a-2) = 0.$

Therefore, $G(2, 2)$; so the radius $r = \overline{GD} = \sqrt{2^2 + 2^2} = 2\sqrt{2}$. ∎

3.5 The Cartesian Equation of Circle　圆的笛卡尔方程

Definition 定义

The **locus** (plural: **loci**) is a set of all points satisfy some property that has been specified.

轨迹是满足特定属性的点的集合.

The locus of equidistance r from any given point C in a plane is a circle, where C is known as the center of the circle, and the radius of the circle is r.

与一个平面上的点 C 距离为 r 的所有点的轨迹构成一个圆,其中,C 是圆的圆心,r 是圆的半径.

The circle having center C and radius r is the set of all points in the plane that are at a distance r from the point C.

Suppose $C(h, k)$, so the distance from any point $P(x, y)$ to C is $\sqrt{(x-h)^2 + (y-k)^2}$; hence the equation of the circle with radius r would be $\sqrt{(x-h)^2 + (y-k)^2} = r$.

A simpler form of the equation of a circle is obtained by squaring both sides of the equation above.

3.6 Standard Equation of a Circle　圆的标准方程

The circle with center (h, k) and radius $r > 0$ has equation $(x-h)^2 + (y-k)^2 = r^2$.

If the circle has center at the origin, then we have $x^2 + y^2 = r^2$.

一个圆的圆心为 (h, k),并且半径 $r > 0$,那么,可以得出方程 $(x-h)^2 + (y-k)^2 = r^2$.

如果该圆的圆心位于坐标轴的原点,那么,可以得出方程 $x^2 + y^2 = r^2$.

Example 例 3.14

Find an equation of the following circles:

(a) center at $(2, 3)$ and diameter of 10;
(b) center at $(1, 2)$ and radius of 3;
(c) center at $(-1, 2)$ and radius of 3;
(d) center at $(-1, -2)$ and radius of 3.

Solution 解

(a) An equation for the circle would be $(x-2)^2 + (y-3)^2 = 5^2$.

(b) An equation for the circle would be $(x-1)^2 + (y-2)^2 = 3^2$.

(c) An equation for the circle would be $(x+1)^2 + (y-2)^2 = 3^2$.

(d) An equation for the circle would be $(x+1)^2 + (y+2)^2 = 3^2$. ■

Since we know the standard equation for a circle center at (h, k) and radius $r > 0$ $(x-h)^2 + (y-k)^2 = r^2$. Let's expand the equation, we have $x^2 - 2hx + h^2 + y^2 - 2ky + k^2 = r^2$; after simplifying we can have a general equation looks like $x^2 + y^2 + 2ax + 2by + c = 0$.

Note 注意

There are other types of conic namely ellipse, hyperbola and parabola which can also be written in the general form as $ax^2 + by^2 + 2cx + 2dy + e = 0$. When $ab > 0$ and $a = b$, it represents a circle.

When $ab > 0$ and $a \neq b$, it represents an ellipse. When $ab < 0$, it represents a hyperbola.

When $ab = 0$, it represents a parabola. We will not discuss ellipse and hyperbola in this course. It is for students' interests to read materials about these topics.

Example 例 ⟨3.15⟩

Find the center and radius of the following circles:

(a) $x^2 + x + y^2 + 4y + 2 = 0$;

(b) $x^2 - 9x + y^2 + 6y - 4 = 0$.

Solution 解

Neither of these equations are in standard form and so to determine the center and radius of the circle we will need to convert them into standard form first.

The process is called completing the square which we have learned in Chapter 1.

(a) $x^2 + x + y^2 + 4y + 2 = 0 \Rightarrow x^2 + x + \dfrac{1}{4} + y^2 + 4y + (2+2) = \dfrac{1}{4} + 2$

$$\Rightarrow \left(x + \dfrac{1}{2}\right)^2 + (y+2)^2 = \dfrac{9}{4}.$$

Hence, the center of the circle is $\left(-\dfrac{1}{2}, -2\right)$ with radius $\dfrac{3}{2}$.

(b) $x^2 - 9x + y^2 + 6y - 4 = 0 \Rightarrow x^2 - 9x + \dfrac{81}{4} + y^2 + 6y + 9 - 4 = \dfrac{81}{4} + 9$

$$\Rightarrow \left(x - \dfrac{9}{2}\right)^2 + (y+3)^2 = \dfrac{81}{4} + 9 + 4 = \dfrac{133}{4}.$$

Hence, the center of the circle is $\left(\dfrac{9}{2}, -3\right)$ with radius $\dfrac{\sqrt{133}}{2}$. ■

Example 例 ⟨3.16⟩

Let $\overline{AB}$ be the diameter of a circle, where $A(3, 3)$ and $B(-5, 1)$. Find the equation of the circle.

Solution 解

Since $\overline{AB}$ is the diameter of the circle, so the center would be the mid-point of $\overline{AB}$. Therefore, the center would have coordinate $\left(\dfrac{3+(-5)}{2}, \dfrac{3+1}{2}\right) \Rightarrow (-1, 2)$.

Hence, the equation of the circle would be $(x+1)^2 + (y-2)^2 = r^2$.

The length of the diameter is $\sqrt{(3+5)^2 + (3-1)^2} = \sqrt{68} = 2\sqrt{17}$, so radius $r = \dfrac{2\sqrt{17}}{2} = \sqrt{17}$.

An equation of the circle could be $(x+1)^2 + (y-2)^2 = 17$. ∎

Example 例 3.17

Let three points $P(1, 2)$, $Q(3, -4)$ and $R(-1, 0)$ be on a circle. Find the equation of the circle.

Solution 解

By Theorem 3.4, we know the perpendicular bisector of the chord would pass through the center of the circle. Therefore, the first step is to construct two chords $\overline{PQ}$ and $\overline{PR}$.

Let the mid-point of $\overline{PQ}$ and $\overline{PR}$ be M_1 and M_2 respectively.

Therefore, $M_1(2, -1)$ and $M_2(0, 1)$. The gradients for $\overline{PQ}$ and $\overline{PR}$ are -3 and 1 respectively.

The perpendicular bisector of $\overline{PQ}$ is $\dfrac{y+1}{x-2} = \dfrac{1}{3} \Rightarrow y = \dfrac{x}{3} - \dfrac{5}{3}$.

The perpendicular bisector of $\overline{PR}$ is $\dfrac{y-1}{x-0} = -1 \Rightarrow y = -x + 1$.

Hence, the intersection of these two perpendicular bisectors would be the center of the circle.

$\dfrac{x}{3} - \dfrac{5}{3} = -x + 1 \Rightarrow x = 2, y = -1$.

Let $C(2, -1)$ be the center of the circle, so the radius of the circle would be the distance from the center to any points on the circle, this implies $r = \overline{CR} = \sqrt{(2+1)^2 + 1^2} = \sqrt{10}$.

The equation of the circle is $(x-2)^2 + (y+1)^2 = 10$. ∎

3.7 Intersection between Line and Circle 直线与圆相交

Finding the intersection between line and circle is similar to finding the intersection between line and parabola. We simply replaced the y term in the circle by the equation of the line $y = ax + b$ to obtain a quadratic equation in terms of x and solve for the resulting equation. As we discuss in Chapter 1, we can use the discriminant D to find the types of intersection. We have three cases.

求直线与圆的交点类似于求直线与抛物线的交点. 只需用直线方程 $y = ax + b$ 代替圆中的 y 项, 得到一

个关于 x 的二次方程，并求解该方程。正如在第 1 章中所讨论的，可以用判别式 D 来求交点的类型，可以得出以下 3 种情况：

（1）Intersect at two distinct points ($D > 0$, Fig. 3.21).

在两个不同的点相交（$D > 0$，图 3.21）。

（2）Intersect at a single point (The line is tangent to the circle) ($D = 0$, Fig. 3.22).

在唯一一个点相交（直线与圆相切）（$D = 0$，图 3.22）。

（3）No Intersection ($D < 0$, Fig. 3.23).

不相交（$D < 0$，图 3.23）。

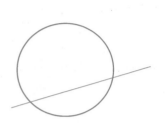
Fig. 3.21

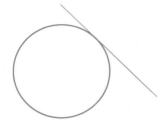
Fig. 3.22

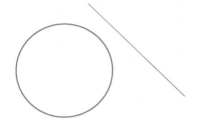
Fig. 3.23

Example 例 3.18

For what value of r does the line $y = 2x - 1$ intersect the circle $(x-1)^2 + (y+2)^2 = r^2$?

Solution 解

Method 1（Using Discriminant）

$(x-1)^2 + (2x-1+2)^2 - r^2 = 0 \Rightarrow 5x^2 + 2x + (2-r^2) = 0$.

We need $D = 2^2 - 4 \times 5 \times (2-r^2) \geq 0 \Rightarrow 20r^2 \geq 36 \Rightarrow r^2 \geq \dfrac{9}{5}$.

So $r \geq \dfrac{3}{\sqrt{5}}$ or $r \leq \dfrac{-3}{\sqrt{5}}$.

Method 2（Using Distance）

The center of the circle is $(1, -2)$, the shortest distance from the point to the line $2x - y - 1 = 0$ is $d = \dfrac{|2 \times 1 - (-2) - 1|}{\sqrt{2^2 + 1^2}} = \dfrac{3}{\sqrt{5}}$. So we need radius $\geq \dfrac{3}{\sqrt{5}}$; hence, $r \geq \dfrac{3}{\sqrt{5}}$ or $r \leq \dfrac{-3}{\sqrt{5}}$. ∎

*3.8 Intersection between Two Circles 圆与圆相交

Given two circles with equations $C_1: (x-a)^2 + (y-b)^2 = r_1^2$ and $C_2: (x-c)^2 + (y-d)^2 = r_2^2$. We know the centers for C_1 and C_2 are (a, b) and (c, d) respectively. So the distance between two centers are $D = \sqrt{(c-a)^2 + (d-b)^2}$; in order to have two distinct intersection between circles we need $r_0 + r_1 > D$ and $D > |r_0 - r_1|$.

If $D < |r_0 - r_1|$, then there are no intersections because one circle is contained within the other.

If $D = r_0 + r_1$, then the two circles touch at a single point (Fig. 3.24). Let the intersection points of the two circles be $A(x_1, y_1)$ and $B(x_2, y_2)$, the line segment $\overline{AB}$ would passing through the two centers of the circles at 90 degrees; and the line segment passing through the two centers has a gradient of $\dfrac{d-b}{c-a}$ with equation $y = \dfrac{d-b}{c-a}x - \dfrac{a(d-b)}{c-a} + b$.

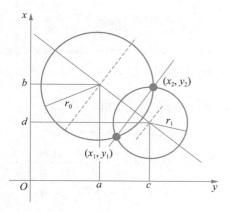

Fig. 3.24

Hence, the line segment $\overline{AB}$ has gradient of $\dfrac{a-c}{d-b}$ with equation

$$y = \dfrac{a-c}{b-d}x - \dfrac{(r_0^2 - r_1^2) + (c^2 - a^2) + (d^2 - b^2)}{2(d-b)}.$$

As discussed in previous section, we know how to find the intersections between lines and circles; therefore the intersection points are easily obtained.

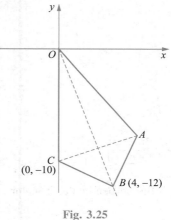

Fig. 3.25

Example 例 3.19

The diagram shows a kite $OABC$ in which OB is the line of symmetry. The coordinates of B and C are $(4, -12)$ and $(0, -10)$ respectively and O is the origin (Fig. 3.25).

(a) Find the equations of OB and AC.

(b) Find, by calculation, the coordinates of A.

Solution 解

(a) The gradient for $\overline{OB}$ is $\dfrac{-12 - 0}{4 - 0} = -3$, hence; the equation for $\overline{OB}$ is

$$y = -3x.$$

It is given that $OABC$ is a kite, so the diagonals intersect at right angle. Therefore, the gradient for $\overline{AC}$ is $\dfrac{-1}{-3} = \dfrac{1}{3}$.

The equation for $\overline{AC}$ is $\dfrac{y - (-10)}{x - 0} = \dfrac{1}{3} \Rightarrow y = \dfrac{1}{3}x - 10$.

(b) We first find the intersection point of $\overline{OB}$ and $\overline{AC}$ $\Rightarrow -3x = \dfrac{1}{3}x - 10 \Rightarrow \dfrac{10x}{3} = 10 \Rightarrow x = 3$.

$(3, -9)$ is the point of intersection and this point is the midpoint of $\overline{AC}$.

Suppose $A(x_A, y_A)$, by the definition of midpoint; we know $\dfrac{x_A + 0}{2} = 3$ and $\dfrac{y_A + (-10)}{2} = -9$.

So $x_A = 6$ and $y_A = -8$, so the coordinate of A is $(6, -8)$.

Example 例 ⟨3.20⟩

Points A and B have coordinates $(a, 2a+1)$ and $(3a-6, 4a)$, respectively. The equation of the perpendicular bisector of AB is $4x - y = 2b$. Find the values of the constants a and b.

Solution 解

$4x - y = 2b \Rightarrow y = 4x - 2b$.

The gradient of $\overline{AB}$ is $\dfrac{4a - (2a+1)}{3a - 6 - a} = \dfrac{2a - 1}{2a - 6} = -\dfrac{1}{4} \Rightarrow a = 1$.

Hence, $A(1, 3)$ and $B(-3, 4)$.

So the midpoint of $\overline{AB}\left(-1, \dfrac{7}{2}\right)$.

This implies that $4 \times (-1) - \dfrac{7}{2} = 2b \Rightarrow b = -\dfrac{15}{4}$. ■

Example 例 ⟨3.21⟩

The coordinates of points A and B are $(2a+1, a-2)$ and $(-a-2, 2a-1)$ respectively, where a is a constant ($a \neq -1$).

(a) Find the gradient of AB.

(b) Find the equation of the perpendicular bisector of AB.

Solution 解

(a) The gradient of $\overline{AB}$ is $\dfrac{2a - 1 - (a - 2)}{-a - 2 - (2a + 1)} = \dfrac{a + 1}{-3a - 3} = \dfrac{a + 1}{-3(a + 1)} = -\dfrac{1}{3}$.

(b) The midpoint M of $\overline{AB}$ is $\left(\dfrac{2a + 1 - a - 2}{2}, \dfrac{a - 2 + 2a - 1}{2}\right) \Rightarrow M\left(\dfrac{a - 1}{2}, \dfrac{3a - 3}{2}\right)$.

The gradient of the perpendicular bisector is $\dfrac{-1}{-\dfrac{1}{3}} = 3$, hence the equation of the perpendicular bisector is

$\dfrac{\left(y - \dfrac{3a - 3}{2}\right)}{x - \left(\dfrac{a - 1}{2}\right)} = 3 \Rightarrow \dfrac{2y - 3a + 3}{2x - a + 1} = 3 \Rightarrow 2y - 3a + 3 = 6x - 3a + 3 \Rightarrow y = 3x$. ■

Example 例 ⟨3.22⟩

The points $A(0, 3)$ and $B(16, 1)$ lie on the curve $y = 4x^2 - 24x + 24$.

(a) Find the equation of the perpendicular bisector of AB.

(b) The perpendicular bisector of AB meets the curve at C and D. Find the distance CD.

Solution 解

(a) The midpoint M of $\overline{AB}$ is $\left(\dfrac{0 + 16}{2}, \dfrac{1 + 3}{2}\right) \Rightarrow (8, 2)$.

The gradient of $\overline{AB}$ is $\dfrac{1-3}{16-0} = -\dfrac{1}{8}$.

The perpendicular bisector of $\overline{AB}$ has gradient of 8.

The equation of the perpendicular bisector of $\overline{AB}$ is $y + 8 = 8(x - 2) \Rightarrow y = 8x - 24$.

(b) $4x^2 - 24x + 24 = 8x - 24 \Rightarrow 4x^2 - 32x + 48 = 0 \Rightarrow x^2 - 8x + 12 = 0 \Rightarrow x = 2$ or 6.

When $x = 2$, $y = 8 \times 2 - 24 = -8$.

When $x = 6$, $y = 8 \times 6 - 24 = 24$.

Hence the coordinates of C and D is $(2, -8)$ $(6, 24)$.

So the distance of CD is $\sqrt{(6-2)^2 + (24+8)^2} = 4\sqrt{65}$.

Example (例) 3.23

The point A has coordinates $(-2, -8)$. The equation of the perpendicular bisector of the line AB is $y = 2x + 6$.

(a) Find the equation of AB.

(b) Find the coordinates of B.

Solution 解

(a) Since the equation of the perpendicular bisector of AB is $y = 2x + 6$.

Hence, the gradient of AB is $-\dfrac{1}{2}$.

So the equation of AB is $y + 8 = -\dfrac{1}{2}(x + 2) \Rightarrow y = -\dfrac{1}{2}x - 9$.

(b) $2x + 6 = -\dfrac{1}{2}x - 9 \Rightarrow x = -6$, so $y = 2 \times (-6) + 6 = -6$.

The midpoint of AB is $(-6, -6)$.

Hence, the coordinates of B is $(-10, -4)$.

Example (例) 3.24

C is the midpoint of the line joining $A(5, 2)$ to $B(3, 8)$. The line through C perpendicular to AB crosses the y-axis at D. Find the equation of the line CD and the distance CD.

Solution 解

The midpoint C is $\left(\dfrac{5+3}{2}, \dfrac{2+8}{2}\right) = (4, 5)$.

The gradient of $\overline{AB}$ is $\dfrac{8-2}{3-5} = -3$.

The perpendicular bisector of $\overline{AB}$ has gradient of $\dfrac{1}{3}$.

The equation of the perpendicular bisector is $y - 5 = \dfrac{1}{3}(x - 4) \Rightarrow y = \dfrac{1}{3}x + \dfrac{11}{3}$.

The coordiantes of D is $\left(0, \dfrac{11}{3}\right)$.

Hence, the distances of $\overline{CD}$ is $\sqrt{(4-0)^2 + \left(5 - \dfrac{11}{3}\right)^2} = \dfrac{4}{3}\sqrt{10}$. ∎

Example 例 ⟨3.25⟩

The line $\dfrac{x}{p} + \dfrac{y}{q} = 2$, where p and q are positive constants, intersects the x- and y-axes at the points A and B respectively. The midpoint of AB lies on the line $3x + 2y = 18$ and the distance $AB = 10$. Find the values of p and q.

Solution 解

When $y = 0$, $x = 2p \Rightarrow A(2p, 0)$.
When $x = 0$, $y = 2q \Rightarrow B(0, 2q)$.
The midpoint of $\overline{AB}$ is (p, q).
So $3p + 2q = 18 \Rightarrow q = \dfrac{18 - 3p}{2}$.

The distance $\overline{AB} = \sqrt{4p^2 + 4q^2} = 10 \Rightarrow p^2 + q^2 = 25 \Rightarrow p^2 + \left(\dfrac{18 - 3p}{2}\right)^2 = 25 \Rightarrow p = 4$ or $\dfrac{56}{13}$, $q = 3$ or $\dfrac{33}{13}$. ∎

Example 例 ⟨3.26⟩

B is the midpoint of AC and A has coordinates $(-2, a)$ and B has coordinates $(b, 3)$, where a and b are constants.

(a) Find the coordinates of C in terms of a and b.

The equation of the perpendicular bisector of the line AC is $y = -\dfrac{1}{3}x + 4$.

(b) Find the values of a and b.

Solution 解

(a) $C(2b + 2, 2 \times 3 - a) \Rightarrow C(2b + 2, 6 - a)$.

(b) B is the midpoint of $\overline{AC}$, and the equation of the perpandicular bisector of $\overline{AC}$ is $y = -\dfrac{1}{3}x + 4$.

So $3 = -\dfrac{1}{3} \times b + 4 \Rightarrow b = 3$, the gradient of $\overline{AC} = \dfrac{3 - a}{3 + 2} = 3 \Rightarrow a = -12$. ∎

Example 例 ⟨3.27⟩

Three points have coordinates $A(1, 6)$, $B(9, 8)$ and $C(a, 5a)$. Find the value of the constant a for which

(a) C lies on the perpendicular bisector of AB;
(b) C lies on the line that passes through A and B.

Solution 解

(a) The midpoint M of $\overline{AB}$ is $\left(\dfrac{1 + 9}{2}, \dfrac{6 + 8}{2}\right) \Rightarrow (5, 7)$.

The gradient of $\overline{AB}$ is $\dfrac{8-6}{9-1} = \dfrac{1}{4}$.

The perpendicular bisector of $\overline{AB}$ has gradient of -4.

The equation of the perpendicular bisector of $\overline{AB}$ is $y - 7 = -4(x - 5) \Rightarrow y = -4x + 27a$.
So $5a = -4a + 27 \Rightarrow a = 3$.

(b) The equation of $\overline{AB}$ is $y - 6 = \dfrac{1}{4}(x - 1) \Rightarrow y = \dfrac{1}{4}x + \dfrac{23}{4}$.

So $5a = \dfrac{1}{4}a + \dfrac{23}{4} \Rightarrow a = \dfrac{23}{19}$.

Example 例 ⟨3.28⟩

Points A and B have coordinates $A(4, 2)$ and $B(6, 16)$.
(a) Find the equation of the perpendicular bisector of AB.
(b) The line through $C(-4, -4)$ parallel to AB meets the perpendicular bisector of AB at the point D. Find the distance BD.

Solution 解

(a) The midpoint M of $\overline{AB}$ is $\left(\dfrac{4+6}{2}, \dfrac{2+16}{2}\right) \Rightarrow (5, 9)$.

The gradient of $\overline{AB}$ is $\dfrac{16-2}{6-4} = 7$.

The perpendicular bisector of $\overline{AB}$ has gradient of $-\dfrac{1}{7}$.

The equation of the perpendicular bisector of $\overline{AB}$ is $y - 2 = -\dfrac{1}{7}(x - 4) \Rightarrow y = -\dfrac{x}{7} + \dfrac{18}{7}$.

(b) The equation of $\overline{AB}$ is $y + 4 = 7(x + 4) \Rightarrow y = 7x + 24$.

$7x + 24 = -\dfrac{1}{7}x + \dfrac{68}{7} \Rightarrow x = -2$.

$y = 7 \times (-2) + 24 = 10$.

Hence, $D(-2, 10)$.

The distance $BD = \sqrt{(6+2)^2 + (10-16)^2} = 10$.

Example 例 ⟨3.29⟩

Points A, B and C have coordinates $A(-3, 1)$, $B(1, 7)$ and $C(2a, 3)$, where a is a constant.
(a) Given that $AB = BC$, calculate the possible values of a.
   The perpendicular bisector of AB intersects the y-axis at D.
(b) Calculate the coordinates of D.

Solution 解

(a) $AB = BC \Rightarrow \sqrt{(-3-1)^2 + (1-7)^2} = \sqrt{(1-2a)^2 + (7-3)^2} \Rightarrow (1-2a)^2 = 16 + 36 - 16 = 36$

$\Rightarrow 1 - 2a = \pm 6 \Rightarrow a = -\dfrac{5}{2}$ or $\dfrac{7}{2}$.

(b) The midpoint M of $\overline{AB}$ is $\left(\dfrac{-3+1}{2}, \dfrac{1+7}{2}\right) \Rightarrow (-1, 4)$.

The gradient of $\overline{AB}$ is $\dfrac{7-1}{1+3} = \dfrac{3}{2}$.

The perpendicular bisector of $\overline{AB}$ has gradient of $-\dfrac{2}{3}$.

The equation of the perpendicular bisector of $\overline{AB}$ is $y - 4 = -\dfrac{2}{3}(x+1) \Rightarrow y = -\dfrac{2}{3}x + \dfrac{10}{3}$.

So, $D\left(0, \dfrac{10}{3}\right)$. ∎

Example 例 3.30

The line with gradient -3 passing through the point $M(m, 2m)$ intersects the x-axis at A and the y-axis at B.

(a) Find the area of triangle AOB in terms of m.

The line through M perpendicular to AB intersects the x-axis at C.

(b) Find the equation of line MC.

Solution 解

(a) The equation of $\overline{AB}$ is $y - 2m = -3(x - m) \Rightarrow y = -3x + 5m$.

So, $A\left(\dfrac{5}{3}m, 0\right)$, $B(0, 5m)$.

The area of triangle $AOB = \dfrac{1}{2} \times OA \times OB = \dfrac{1}{2} \times \dfrac{5}{3}m \times 5m = \dfrac{25}{6}m^2$.

(b) The line through M is perpendicular to AB.

So the gradient of this line is $\dfrac{1}{3}$.

The equation of line MC: $y - 2m = \dfrac{1}{3}(x - m) \Rightarrow y = \dfrac{1}{3}x + \dfrac{5}{3}m$. ∎

Example 例 3.31

Two points have coordinates $A(2, 3)$, $B(10, 7)$. The third point C lies on the perpendicular bisector of the line AB. C also lies on the line parallel to AB through $(2, 8)$.

(a) Find the equation of the perpendicular bisector of AB.

(b) Calculate the coordinates of C.

Solution 解

(a) The midpoint M of $\overline{AB}$ is $\left(\dfrac{2+10}{2}, \dfrac{3+7}{2}\right) \Rightarrow (6, 5)$.

The gradient of $\overline{AB}$ is $\dfrac{7-3}{10-2} = \dfrac{1}{2}$.

The perpendicular bisector of $\overline{AB}$ has gradient of -2.

The equation of the perpendicular bisector of $\overline{AB}$ is $y - 5 = -2(x - 6) \Rightarrow y = -2x + 17$.

(b) The equation of the line parallel to AB is $y - 8 = \frac{1}{2}(x - 2) \Rightarrow y = \frac{1}{2}x + 7$.

$\frac{1}{2}x + 7 = -2x + 17 \Rightarrow x = 4$.

$y = \frac{1}{2} \times 4 + 7 = 9$.

So, the coordinates of C is $(4, 9)$.

Example 3.32

The point A has coordinates $(a, 1)$ and the point B has coordinates $(5, 5a + 2)$, where a is a constant.
(a) The distance of AB is 26, find the possible values of a.
(b) Line with equation $x + 5y = 10$ is perpendicular to AB, find the value of a.

Solution

(a) $AB = \sqrt{(5-a)^2 + (5a + 2 - 1)^2} = 26 \Rightarrow 25 - 10a + a^2 + 25a^2 + 10a + 1 = 676$
$\Rightarrow 26a^2 = 650 \Rightarrow a = \pm 5$.

(b) $x + 5y = 10 \Rightarrow y = -\frac{1}{5}x + 10$.

The gradient of AB is $\frac{5a + 2 - 1}{5 - a} = 5 \Rightarrow a = \frac{12}{5}$.

Example 3.33

The diagram below shows a parallelogram $ABCD$ in which AB is parallel to DC. The coordinates of A, B and C are $(3, 9)$, $(1, 2)$ and $(7, 1)$ respectively (Fig. 3.26).
(a) Find the equation of AD.
(b) Find, by calculation, the coordinates of D.

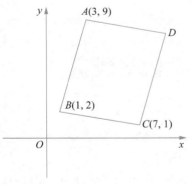

Fig. 3.26

Solution

(a) The gradient of $BC = \frac{1 - 2}{7 - 1} = -\frac{1}{6}$.

AD is parallel to BC.

So, the gradient of AD is also $-\frac{1}{6}$.

The equation of AD is $y - 9 = -\frac{1}{6}(x - 3) \Rightarrow y = -\frac{1}{6}x + \frac{19}{2}$.

(b) The gradient of $AB = \frac{9 - 2}{3 - 1} = \frac{7}{2}$.

AB is parallel to CD.

So, the gradient of CD is also $\dfrac{7}{2}$.

The equation of CD is $y - 1 = \dfrac{7}{2}(x - 7) \Rightarrow y = \dfrac{7}{2}x - \dfrac{47}{2}$, $-\dfrac{1}{6}x + \dfrac{19}{2} = \dfrac{7}{2}x - \dfrac{47}{2} \Rightarrow x = 9$.

Hence, $y = \dfrac{7}{2} \times 9 - \dfrac{47}{2} = 8$.

So, the coordinates of D is $(9, 8)$. ∎

The point A has coordinates $(k + 2, 2k)$ and the point B has coordinates $(4, 3k - 2)$, where k is a constant.

(a) Find the gradient of a line perpendicular to AB.

(b) Given that the distance AB is $8\sqrt{2}$, find the possible values of k.

(a) The gradient of $\overline{AB}$ is $\dfrac{3k - 2 - 2k}{4 - k - 2} = \dfrac{k - 2}{2 - k} = -1$.

The line of perpendicular to $\overline{AB}$ has gradient of 1.

(b) The distance $AB = \sqrt{(k + 2 - 4)^2} + \sqrt{(2k + 3k - 2)^2} = 8\sqrt{2} \Rightarrow 2(k - 2)^2 = 128 \Rightarrow k - 2 = \pm 8 \Rightarrow k = 10$ or -6. ∎

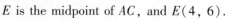

The diagram below shows a parallelogram $ABCD$, in which the equation of AB is $y = -x + 4$ and the equation of AD is $y = \dfrac{1}{2}x + 1$.

The diagonals AC and BD meet at the point $E(4, 6)$. Find the coordinates of A, B, C and D (Fig. 3.27).

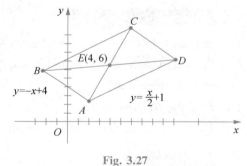

Fig. 3.27

Solution

$-x + 4 = \dfrac{x}{2} + 1 \Rightarrow x = 2, y = \dfrac{2}{2} + 1 = 2 \Rightarrow A(2, 2)$.

Since $ABCD$ is a parallelogram, so their diagonals bisect each other.

E is the midpoint of AC, and $E(4, 6)$.

Hence, $C(6, 10)$, $AB \mathbin{/\mkern-5mu/} CD$, $AD \mathbin{/\mkern-5mu/} BC$.

The equation of BC: $y - 10 = \dfrac{1}{2}(x - 6) \Rightarrow y = \dfrac{x}{2} + 7$, $\dfrac{x}{2} + 7 = -x + 4 \Rightarrow x = -2, y = 2 + 4 = 6 \Rightarrow B(-2, 6)$.

The equation of CD: $y - 10 = -(x - 6) \Rightarrow y = -x + 16$, $-x + 16 = \dfrac{x}{2} + 1 \Rightarrow x = 10, y = -10 + 16 = 6 \Rightarrow D(10, 6)$. ∎

Example 3.36

The point P is the reflection of the point $(2, 4)$ in the line $y + 2x = 18$. Find the coordinates of P.

Solution

$y + 2x = 18 \Rightarrow y = -2x + 18$.

The gradient of the line through point P is $\frac{1}{2}$, and the equation of it is $y - 4 = \frac{1}{2}(x - 2) \Rightarrow y = \frac{1}{2}x + 3$.

$\frac{1}{2}x + 3 = -2x + 18 \Rightarrow x = 6, y = -2 \times 6 + 18 = 6 \Rightarrow (6, 6)$.

$(6, 6)$ is midpoint of P and $(2, 4)$.

So, the coordinates of P is $(10, 8)$.

Example 3.37

The circle $x^2 + y^2 + 8x - 6y = 0$ has centre C and passes through points A and B.

(a) State the coordinates of C.

(b) Point $D(-5, 1)$ is midpoint of AB. Find the equation of AB and find the x-coordinates of A and B by calculation.

Solution

(a) $x^2 + y^2 + 8x - 6y = 0 \Rightarrow (x + 4)^2 + (y - 3)^2 = 25$.

Hence, the center $C(-4, 3)$.

(b) Since D is midpoint of AB, C is the center of the circle.

So, CD is perpendicular to AB.

The gradient of CD is $\frac{1 - 3}{-5 + 4} = 2$,

Hence, the gradient of AB is $-\frac{1}{2}$.

The equation of AB is $y - 1 = -\frac{1}{2}(x + 5) \Rightarrow y = -\frac{1}{2}x - \frac{3}{2}$.

$(x + 4)^2 + \left(-\frac{1}{2}x - \frac{3}{2} - 3\right)^2 = 25 \Rightarrow x^2 + 8x + 16 + \frac{1}{4}x^2 + \frac{9}{2}x + \frac{81}{4} = 25 \Rightarrow x^2 + 10x + 9 = 0 \Rightarrow x = -1 \text{ or } -9$.

Summary of Key Theories　核心定义总结

(1) The gradient between two points $A(x_1, y_1)$ and $B(x_2, y_2)$ is $m = \dfrac{y_2 - y_1}{x_2 - x_1}$.

The equation of the line segment $\overline{AB}$ is given by $\dfrac{y-y_1}{x-x_1}=\dfrac{y_2-y_1}{x_2-x_1}$ or $\dfrac{y-y_2}{x-x_2}=\dfrac{y_2-y_1}{x_2-x_1}$.

The x-intercept of the line is determined when $y = 0$.

The y-intercept of the line is determined when $x = 0$.

(2) Given two lines $l_1: y = m_1 x + b_1$ and $l_2: y = m_2 x + b_2$.

If $m_1 = m_2$ and $b_1 \neq b_2$, then two lines are parallel.

If $m_1 \cdot m_2 = -1$ regardless the values of b_i; then two lines are perpendicular.

(3) A tangent to a circle is a line that just touches the circle at one point.

(4) The **locus** (plural: **loci**) is a set of all points satisfy some property that has been specified.

(5) The locus of equidistance r from any given point C in a plane is a circle, where C is known as the center of the circle, and the radius of the circle is r.

(6) The circle with center (h, k) and radius $r > 0$ has equation $(x-h)^2 + (y-k)^2 = r^2$. If the circle has center at the origin, then we have $x^2 + y^2 = r^2$.

Circular Measure

第 4 章 弧 度 法

We should be familiar with angles measured in degrees; however, degree is not the only unit for angles. A useful angle measure other than degree is the radian measure.

> **Definition** 定义
> If the length of an arc is equal to the radius of the circle subtended at an angle θ, then the angle θ is said to have a measure of 1 radian (Fig. 4.1).
> 如果一段弧的长度等于圆的半径,那么,该弧对应的角 θ 为 1 弧度(图 4.1).

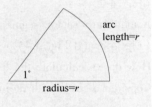

Fig. 4.1

As we know, the circumference of a circle with radius r is $2\pi r$; and a circle is corresponding to $360°$. Therefore, the conversion between degree measure and radian measure would be:
$360° \equiv 2\pi$ radians $\Rightarrow 180° \equiv \pi$ radians.
A more general formula would be:
$x° = \dfrac{x°}{180°} \times \pi$ radians or $x° = \dfrac{x°}{360°} \times 2\pi$ radians, y radians $= \dfrac{y}{\pi} \times 180°$ or y radians $= \dfrac{y}{2\pi} \times 360°$.
The following table shows some special angles measured in degrees and radians.

Tab. 4.1

Degrees	0°	±30°	±45°	±60°	±90°	±180°	±360°
Radians	0	$\pm\dfrac{\pi}{6}$	$\pm\dfrac{\pi}{4}$	$\pm\dfrac{\pi}{3}$	$\pm\dfrac{\pi}{2}$	$\pm\pi$	$\pm 2\pi$

> **Note** 注意
> We will discuss negative angles in next chapter, and we usually use rad as an abbreviation for radians.

Convert the following angles from degrees to radians and vice versa.

(a) 5°; (b) 32°; (c) 197°; (d) 0.37 rad; (e) 1.62 rad; (f) 2.19 rad.

(a) $\dfrac{5°}{180°} \times \pi = \dfrac{\pi}{36}$ rad; (b) $\dfrac{32°}{180°} \times \pi = \dfrac{8\pi}{45}$ rad; (c) $\dfrac{197°}{180°} \times \pi = \dfrac{197\pi}{180}$ rad;

(d) $\dfrac{0.37}{\pi} \times 180° = 21.2°$; (e) $\dfrac{1.62}{\pi} \times 180° = 92.8°$; (f) $\dfrac{2.19}{\pi} \times 180° = 125.5°$.

4.1 Arc Length 弧长

We should be familiar with the terms relating to a circle:
A minor arc/sector/segment means it involves less than half of a circle (Fig. 4.2).
A major arc/sector/segment means it involves more than half of a circle (Fig. 4.3 and Fig. 4.4).
How do we calculate the length of an arc?
Consider a circle with radius r, we know the circumference of a circle is $2\pi r$. Suppose the sector with angle $\theta°$, we now use ratio to find the arc length l (Fig. 4.5).

$l = 2\pi r \times \dfrac{\theta°}{360°} = r \times \dfrac{\theta°}{360°} \times 2\pi = r \times \theta_{\text{rad}}$.

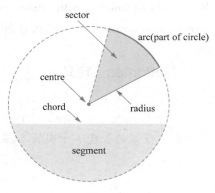

Fig. 4.2

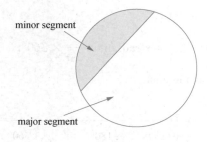

Fig. 4.3

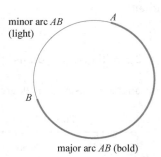

Fig. 4.4

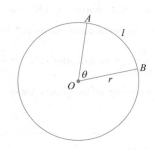

Fig. 4.5

Theorem 4.1 定理 4.1

The arc length l of a sector with angle θ radians is determined by $l = r\theta$, where r is the radius of the circle.
一个扇形的角弧度为 θ，该扇形的弧长为 $l = r\theta$，其中，r 为圆的半径。

Proof 证明

It is proved above.

Example 例 4.2

(a) Find the arc length of a sector of a circle with radius 12 cm and angle 2.3 radians.
(b) Find the arc length of a sector of a circle with radius 10 cm and angle 64°.

Solution 解

(a) Since the angle is given in radian, so we can apply Theorem 4.1 directly.
$l = r\theta \Rightarrow l = 12 \times 2.3 = 27.6 (\text{cm})$.

(b) The angle is in degree not radian, so we have to do the conversion first.
$64° = \dfrac{64°}{180°} \times \pi \approx 1.117 \text{ rad}$. Therefore, $l = r\theta \Rightarrow l = 10 \times 1.117 \approx 11.2 (\text{cm})$.

Example 例 4.3

(a) Find the radius and perimeter of a sector with arc length of 6 m and an angle of 3 radians.
(b) Find the radius and perimeter of a sector with arc length of 7.92 m and an angle of 132°.

Solution 解

(a) $l = r\theta \Rightarrow 6 = r \times 3 \Rightarrow r = 2 (\text{m})$. The perimeter would be $2r + l = 2 \times 2 + 6 = 10 (\text{m})$.

(b) $132° = \dfrac{132°}{180°} \times \pi \approx 2.30383 \text{ rad}$; $l = r\theta \Rightarrow 7.92 = r \times 2.30383 \Rightarrow r = 3.437746771 \approx 3.44 (\text{m})$.

The perimeter would be $2r + l = 2 \times 3.437746771 + 7.92 \approx 14.8 (\text{m})$.

4.2 Area of Sector 扇形面积

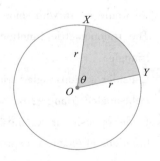

Fig. 4.6

Consider a circle with radius r, we know the area of a circle is πr^2. Suppose the sector with angle $\theta°$, we now use ratio to find the area of the sector (Fig. 4.6).

$$\text{Area} = \pi r^2 \times \dfrac{\theta°}{360°} = r^2 \times \dfrac{\theta°}{2 \times 360°} \times 2\pi = \dfrac{1}{2} r^2 \theta_{\text{rad}}.$$

Theorem 4.2 定理 4.2

The area A of a sector with angle θ radians is determined by $A = \dfrac{1}{2} r^2 \theta$, where r is the radius of the circle.

一个扇形的角弧度为 θ，该扇形的面积为 $A = \dfrac{1}{2} r^2 \theta$，其中，$r$ 为圆的半径。

Proof 证明

It is proved above. ∎

Example 例 4.4

(a) Find the area of a sector of a circle with radius 8 cm and angle 1.32 radians.

(b) Find the area of a sector of a circle with radius 15 cm and angle 36°.

Solution 解

(a) Since the angle is given in radian, so we can apply Theorem 4.2 directly.

$$A = \frac{1}{2}r^2\theta \Rightarrow A = \frac{1}{2} \times 8^2 \times 1.32 = 42.24 \approx 42.2(\text{cm}^2).$$

(b) The angle is in degree not radian, so we have to do the conversion first.

$$36° = \frac{36°}{180°} \times \pi \approx 0.628\,32 \text{ rad}.$$

Therefore, $A = \frac{1}{2}r^2\theta \Rightarrow A = \frac{1}{2} \times 15^2 \times 0.628\,32 = 70.685\,83 \approx 70.7(\text{cm}^2).$ ∎

4.3 Further Problems Involving Arcs and Sectors
   弧与扇形的拓展题目

The majority of the problems in A-level exam involves the use of trigonometry we learned in IGCSE. Therefore, we would begin with some reviews of trigonometry.

The trigonometric functions are often used to express the relationships between the sides and angles of a triangle.

Consider a right-angled triangle as shown, if θ is one of the acute angle in a right-angled triangle; we refer the three sides as adj (side adjacent θ), opp (side opposite θ), and hyp (the hypotenuse). Then, the trigonometric functions of θ can be expressed as the ratios of the sides (Fig. 4.7).

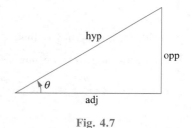

Fig. 4.7

$$\sin\theta = \frac{\text{opp}}{\text{hyp}}, \qquad \cos\theta = \frac{\text{adj}}{\text{hyp}}, \qquad \tan\theta = \frac{\text{opp}}{\text{adj}},$$
$$\csc\theta = \frac{\text{hyp}}{\text{opp}}, \qquad \sec\theta = \frac{\text{hyp}}{\text{adj}}, \qquad \cot\theta = \frac{\text{adj}}{\text{opp}}.$$

Since $\tan\theta = \frac{\text{opp}}{\text{adj}} = \frac{\text{opp/hyp}}{\text{adj/hyp}} = \frac{\sin\theta}{\cos\theta}$ and $\sin^2\theta + \cos^2\theta = \left(\frac{\text{opp}}{\text{hyp}}\right)^2 + \left(\frac{\text{adj}}{\text{hyp}}\right)^2 = \frac{\text{opp}^2 + \text{adj}^2}{\text{hyp}^2} = \frac{\text{hyp}^2}{\text{hyp}^2} = 1.$

So we obtained the two basic trigonometric identities which we should always remember.

Note 注意

In North America, we use $\csc\theta$ instead of $\operatorname{cosec}\theta$.

Theorem 4.3 (The Sine Law)　定理 4.3（正弦定理）

Consider a triangle as shown in Fig. 4.8, the sides are named to correspond to the opposite angles.

Sine Law: $\dfrac{\sin A}{a} = \dfrac{\sin B}{b} = \dfrac{\sin C}{c}$.

如图 4.8 所示，一个三角形的三条边分别与它们的对角对应命名，则有正弦定理 $\dfrac{\sin A}{a} = \dfrac{\sin B}{b} = \dfrac{\sin C}{c}$.

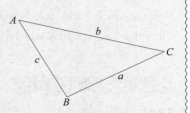

Fig. 4.8

Proof 证明

Let's construct a perpendicular height h from A to BC (Fig. 4.9). Use basic trigonometry $\sin B = \dfrac{h}{c} \Rightarrow h = c\sin B$, $\sin C = \dfrac{h}{b} \Rightarrow h = b\sin C$.

Hence, $c\sin B = b\sin C \Rightarrow \dfrac{\sin B}{b} = \dfrac{\sin C}{c}$. We can draw another perpendicular height from C or B to corresponding sides to obtained the relationship $\dfrac{\sin A}{a} = \dfrac{\sin B}{b} = \dfrac{\sin C}{c}$. ∎

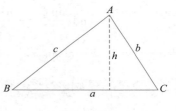

Fig. 4.9

Theorem 4.4 (The Cosine Law)　定理 4.4（余弦定理）

Consider a triangle as shown in Fig. 4.10, the sides are named to correspond to the opposite angles.

Cosine Law:

$a^2 = b^2 + c^2 - 2bc\cos A$　or

$b^2 = a^2 + c^2 - 2ac\cos B$　or

$c^2 = a^2 + b^2 - 2ab\cos C$.

如图 4.10 所示，一个三角形的三条边分别与它们的对角对应命名，则有余弦定理：

$a^2 = b^2 + c^2 - 2bc\cos A$　或

$b^2 = a^2 + c^2 - 2ac\cos B$　或

$c^2 = a^2 + b^2 - 2ab\cos C$.

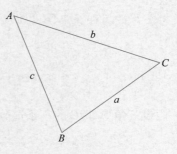

Fig. 4.10

Proof 证明

Let's construct a perpendicular height h from A to BC (Fig. 4.11).

Use Pythagras' Theorem, we obtained $c^2 = h^2 + (a - b\cos C)^2$.

Since $h = b\sin C \Rightarrow c^2 = b^2\sin^2 C + (a - b\cos C)^2 \Rightarrow c^2 = b^2\sin^2 C + (a - b\cos C)^2 = b^2\sin^2 C + a^2 + b^2\cos^2 C - 2ab\cos C$.

Because $\sin^2\theta + \cos^2\theta = 1$, this implies $c^2 = a^2 + b^2 - 2ab\cos C$.

The other two versions can be proved using the same approaches.

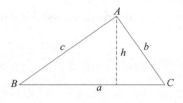

Fig. 4.11

Example 例 4.5

A triangle has sides $a = 4$, $b = 7$ and angle $C = 60°$. Find side c and the angle B and angle A.

Solution 解

Using the Cosine Law: $c^2 = a^2 + b^2 - 2ab\cos C = 4^2 + 7^2 - 2 \times 4 \times 7\cos(60°) = 37$.

So side c equals to $\sqrt{37}$. Now using the Sine Law $\dfrac{\sin A}{a} = \dfrac{\sin B}{b} = \dfrac{\sin C}{c} \Rightarrow \dfrac{\sin A}{a} = \dfrac{\sin C}{c}$ and $\dfrac{\sin B}{b} = \dfrac{\sin C}{c} \Rightarrow \dfrac{\sin A}{4} = \dfrac{\sin 60°}{\sqrt{37}}$ and $\dfrac{\sin B}{7} = \dfrac{\sin 60°}{\sqrt{37}}$.

Hence, angle A is about $34.7°$ and angle B is about $85.3°$.

Example 例 4.6

A triangle has sides $b = 4$, $c = 5$ and angle $B = 30°$. Find side a.

Solution 解

By the Cosine Law, we have
$$b^2 = a^2 + c^2 - 2ac\cos B \Rightarrow 16 = a^2 + 25 - 10a\cos 30° \Rightarrow a^2 - 5\sqrt{3}a + 9 = 0.$$

So $a = \dfrac{5\sqrt{3} \pm \sqrt{75 - 4 \times 9}}{2} = \dfrac{5\sqrt{3} \pm \sqrt{39}}{2}$, $a \approx 7.45$ or $a \approx 1.21$.

Theorem 4.5 (The Area of Triangle)   定理 4.5（三角形面积）

Given triangle ABC and the sides are named to correspond to the opposite angles.

The area of the triangle would be $\dfrac{1}{2}ab\sin C$.

假设三角形 ABC 的三条边分别与它们的对角对应命名，该三角形的面积为 $\dfrac{1}{2}ab\sin C$。

Proof 证明

Consider a general triangle as shown in Fig. 4.12. We constructed the altitude h from A to BC.

We know the area of a triangle is $\frac{1}{2}($ Base $\times$ Height $) = \frac{1}{2}ah$.

Since, $\sin C = \frac{h}{b} \Rightarrow h = b\sin C$.

Hence, Area $= \frac{1}{2}ah = \frac{1}{2}ab\sin C$.

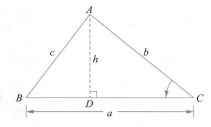

Fig. 4.12

Example 例 4.7

Find the area of the triangle as shown in Fig. 4.13.

Solution 解

Using the area formula：

$A = \frac{1}{2}(10.2)(6.4)\sin\left(\frac{2\pi}{3}\right) \approx 28.3 (\text{cm}^2)$.

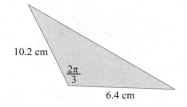

Fig. 4.13

Theorem 4.6 (Heron's Formula)　定理 4.6（海伦公式）

Given triangle ABC with corresponding sides a, b and c respectively. The area of the triangle would be $\sqrt{s(s-a)(s-b)(s-c)}$ where $s = \frac{a+b+c}{2}$.

假如三角形的三条边分别为 a, b 和 c，这个三角形的面积为 $\sqrt{s(s-a)(s-b)(s-c)}$，其中，$s = \frac{a+b+c}{2}$。

Proof 证明

Consider a general triangle as shown in Fig. 4.14. We constructed the altitude h from A to BC.

From the Cosine Law $c^2 = a^2 + b^2 - 2ab\cos C$

$\Rightarrow \cos C = \frac{a^2 + b^2 - c^2}{2ab}$.

Since $\sin^2 C + \cos^2 C = 1 \Rightarrow \sin C = \sqrt{1 - \cos^2 C}$; and we know the area of a triangle is

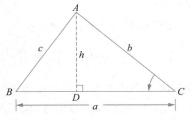

Fig. 4.14

$\frac{1}{2}ab\sin C = \frac{1}{2}ab\sqrt{1-\cos^2 C} = \frac{1}{2}ab\sqrt{1 - \left(\frac{a^2+b^2-c^2}{2ab}\right)^2} = \frac{1}{2}ab\sqrt{\frac{4a^2b^2 - (a^2+b^2-c^2)^2}{4a^2b^2}}$

$= \frac{1}{4}\sqrt{(2ab)^2 - (a^2+b^2-c^2)^2} = \frac{1}{4}\sqrt{(2ab + a^2 + b^2 - c^2)(2ab - a^2 - b^2 + c^2)}$

$= \frac{1}{4}\sqrt{((a+b)^2 - c^2)(c^2 - (a-b)^2)} = \frac{1}{4}\sqrt{(a+b+c)(a+b-c)(c+a-b)(c-a+b)}$.

> Let $s = \dfrac{a+b+c}{2}$, so $s-c = \dfrac{a+b-c}{2}$, $s-b = \dfrac{c+a-b}{2}$ and $s-a = \dfrac{c-a+b}{2}$.
> Then the area of the triangle becomes $\dfrac{1}{4}\sqrt{2s \times 2(s-c) \times 2(s-b) \times 2(s-a)} = \sqrt{s(s-a)(s-b)(s-c)}$. ∎

Example 例 4.8

Find the area of the triangle as shown in Fig. 4.15.

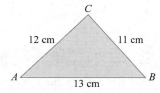

Fig. 4.15

Solution 解

$s = \dfrac{11+12+13}{2} = 18.$

$A = \sqrt{18(18-11)(18-12)(18-13)} = \sqrt{3\,780} \approx 61.5(\text{cm}^2).$ ∎

Example 例 4.9

In this picture (Fig. 4.16), there exists a circle with center A. It's radius is r cm. Angle $ABC = \pi/2$ radians. Angle $BAC = \alpha$. Point A, D and C are on the same line.

(a) Use the terms of r and α to show the area of the shaded region.

(b) For the specific situation as $r=5$ and $\alpha=0.8$, what is the perimeter of the shaded region?

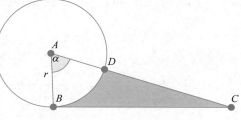

Fig. 4.16

Solution 解

(a) $\dfrac{BC}{AB} = \tan\alpha.$

$BC = AB \times \tan\alpha = r\tan\alpha.$

$S_{ABC} = \dfrac{1}{2}r^2 \tan\alpha,\; S_{ABD} = \dfrac{1}{2}r^2\alpha,\; S_{\text{shaded}} = \dfrac{1}{2}r^2\tan\alpha - \dfrac{1}{2}r^2\alpha.$

(b) $BC = AB \times \tan\alpha = r\tan\alpha = 5\tan 0.8 = 5.148\,19.$

Arc length $= r\alpha = 0.8 \times 5 = 4.$

$\cos\alpha = \dfrac{AB}{AC} \Rightarrow AC = \dfrac{AB}{\cos\alpha} = \dfrac{5}{\cos 0.8} = 7.176\,62.$

$C_{\text{shaded}} = BC + AC + \text{Arc} - \text{radius} = 15.5.$ ∎

Example 例 4.10

The diagram shows two point on the circle with center A. its radius is r cm. Both DC and DB are the tangent line to circle. Angle CAB is α (Fig. 4.17).

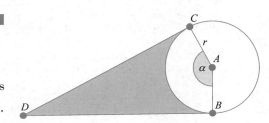

Fig. 4.17

(a) When the area of the sector BAC and shaded region are same, can you show the fact that $\tan\dfrac{\alpha}{2} = \alpha$?

(b) If $r = 10$ cm and the length of the minor arc BC is 30 cm, what is the area of the shaded region?

Solution

(a) CD or $BD = r\tan\dfrac{\alpha}{2}$, $AD = \dfrac{r}{\cos\dfrac{\alpha}{2}}$.

$S_{\text{sector } BAC} = \dfrac{1}{2}r^2\alpha,$

$S_{\text{shaded}} = 2 \times \dfrac{1}{2}r\dfrac{r}{\cos\dfrac{\alpha}{2}}\sin\dfrac{\alpha}{2} - \dfrac{1}{2}r^2\alpha = r^2\tan\dfrac{\alpha}{2} - \dfrac{1}{2}r^2\alpha,$

$S_{\text{shaded}} = S_{\text{sector } BAC} \Rightarrow \tan\dfrac{\alpha}{2} = \alpha.$

(b) $2r\alpha = $ Arc length $\xrightarrow{r = 10, \text{ arc length} = 30} \alpha = 1.5.$

Arc length $= r\alpha$, $r = 10$, arc length $= 30$, $\alpha = \dfrac{30}{10} = 3.$

$S_{\text{shaded}} = r^2\tan\dfrac{\alpha}{2} - \dfrac{1}{2}r^2\alpha = 10^2 \times \tan 1.5 - \dfrac{1}{2} \times 10^2 \times 3 = 1\,260.$

Example 4.11

The picture below is a right triangle ABC with angle ABC is equal to $\dfrac{\pi}{2}$ (Fig. 4.18). $AB = AE = 8$ cm. The length of arc BE is 10 cm. Point A, E and C is on one line. What is the area of shaded region?

Solution

Angle $CAB = \dfrac{BE}{AE} = \dfrac{10}{8} = 1.25.$

$BC = AB \times \tan\alpha = 24.076\,6.$

$S_{ABC} = \dfrac{1}{2}AB \times BC = \dfrac{1}{2} \times 8 \times 24.076\,6 = 96.306\,2.$

$S_{ABE} = \dfrac{1}{2}r^2\alpha = 40.$

$S_{\text{shaded}} = \dfrac{1}{2}r^2\tan\alpha - \dfrac{1}{2}r^2\alpha = 56.3.$

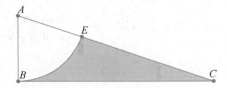

Fig. 4.18

Example 4.12

This diagram shows a triangle ABC in which $AC = 145$ mm, $AB = 25$ mm and $BC = 150$ mm. The circular arcs AD and AE have centres at B and C respectively, where D and E lie on BC (Fig. 4.19).

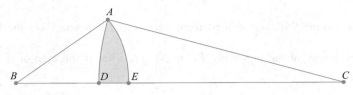

Fig. 4.19

(a) Show that angle $ABC = 1.287$ radians and angle $ACB = 0.166\,3$ radians, correct to 4 decimal places.

(b) Find the area of the shaded region.

Solution

(a) If we draw a point AF which is perpendicular to BC at point F. We have two right triangle ABF and AFC. If we suppose $BF = x$ and $FC = BC - BF = 150 - x$.

With two right triangles with the same BF, we can get
$$145^2 - (150-x)^2 = 25^2 - x^2 \Rightarrow x = 7 \text{ and } AF = 24.$$
$$\sin(ABC) = \frac{AF}{AB} \text{ and } \sin(ACB) = \frac{AF}{AC} \Rightarrow ABC = 1.287 \text{ and } ACB = 0.166\,3.$$

(b) $S_{\text{shaded}} = S_{\text{sector } a} + S_{\text{sector } b} - S_{ABC}$
$$= \frac{1}{2} \times AB^2 \times \text{angle}(ABC) + \frac{1}{2} \times AC^2 \times \text{angle}(ACB) - \frac{1}{2} \times AF \times BC = 350.4.$$

Example 4.13

The diagram shows a semicircle with center C and radius 10 cm. The radius CD is perpendicular to diameter AB. Point B, E, C, A is on the same line. In the meanwhile, DE is an arc of a circle with center A (Fig. 4.20).

(a) Calculate the length of the arc DE.

(b) Find the value of $\dfrac{\text{area of region } X}{\text{area of region } Y}$, correct to 4 decimal places.

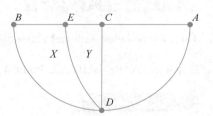

Fig. 4.20

Solution

(a) Connect AD, in the triangle ADC, we can use Pythagoras to get AD. $AD = AE = \dfrac{10}{\cos\dfrac{\pi}{4}} = 10\sqrt{2}$.

Arc $DE = 10\sqrt{2} \times \dfrac{\pi}{4} = 11.1$.

(b) $S_{\text{sector } AED} = \dfrac{1}{2} \times 200 \times \dfrac{\pi}{4} = 25\pi$.

$S_Y = 25\pi - 50$.

$S_X = \dfrac{\pi}{4} \times 10 \times 10 - S_Y = 50$.

ratio $= \dfrac{50}{25\pi - 50} = 1.75$.

Chapter 4　Circular Measure　第四章　弧　度　法

Example 例 ⟨4.14⟩

There is a rectangle $ABCD$ in which $AB = 5$ units and $BC = 13$ units. Points E and F lie on AC. BE is an arc of a circle with center A. BF is an arc of a circle with center D (Fig. 4.21).

(a) Show that angle $BDF = 0.3948$ radians, correct to 4 decimal places.
(b) Calculate the areas of the sectors DBF and ABE.
(c) Calculate the area of the shaded region.

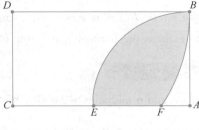

Fig. 4.21

Solution 解

(a) $\sin^{-1}\dfrac{5}{13} = 0.3948$.

(b) $S_{\text{sector } DBF} = \dfrac{1}{2} \times 13 \times 13 \times 0.3948 = 33.3598 \approx 33.4$.

$S_{\text{sector } ABE} = \dfrac{1}{2} \times 5 \times 5 \times \dfrac{\pi}{2} = 19.638 \approx 19.6$.

(c) Connect DF, in the right triangle, $DF = DB = 13$, $DC = 5$. By Pythagoras, we have $CF = 12$.

$S_{\text{sector } DFC} = \dfrac{1}{2} \times 5 \times 12 = 30$.

$S_{\text{shaded}} = S_{\text{sector } DBF} + S_{\text{sector } ABE} - (S_{\text{rect}} - S_{\triangle DFC}) = 33.3598 + 19.638 - (65 - 30) \approx 18.0$. ■

Example 例 ⟨4.15⟩

For this diagram, $ABOC$ is a sector of a circle with center A with radius 13 cm. The length of the chord BC is 10 cm. Points A, E, O are on one line and point E is the midpoint of BC. AE is perpendicular to BC. The shaded region is bounded by the chord BC and by the arc of a circle with center O and radius OC (Fig. 4.22).

(a) Find the angle CBO in radian correct to 4 decimal places.
(b) Find the perimeter of the shaded region.
(c) Solve the area of the shaded region.

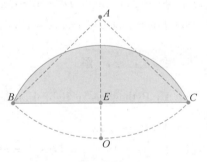

Fig. 4.22

Solution 解

(a) In the right triangle AEC, by Pythagoras, $EC = 1/2 \times 10 = 5$ and $AC = AO = AB = 13$, so we have $AE = 12$ and $EO = 1 \Rightarrow \angle CBO = \angle ECO = \tan^{-1}\left(\dfrac{1}{5}\right) = 0.1974$.

(b) $CO = \sqrt{5^2 + 1^2} = \sqrt{26}$.

Arc $BC = \sqrt{26} \cdot 2\left(\dfrac{\pi}{2} - 0.1974\right) = 15.8177$.

$C_{\text{shaded}} = 15.8177 + 10 = 25.8177 \approx 25.8$.

(c) $S_{\text{shaded}} = S_{\text{sector}} - S_{\triangle BOC} = \dfrac{1}{2} \cdot (\sqrt{26})^2 \cdot 2\left(\dfrac{\pi}{2} - 0.1974\right) - \dfrac{1}{2} \cdot 1 \cdot 10 \approx 30.7$. ■

Example 例 <4.16>

This picture shows a circle with radius r cm and center A. Points B and C are on the circle. $BCDE$ is a rectangle. Angle $BAC = \alpha$ radians. $BE = CD = r$ Use r and α to express the perimeter and area of the shaded region (Fig. 4.23).

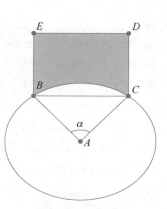

Fig. 4.23

Solution 解

If we add a midpoint M for BC. We will have two right triangles. By Pythagoras, we have $\dfrac{CM}{AC} = \sin\dfrac{\alpha}{2} \Rightarrow BC = 2CM = 2r\sin\dfrac{\alpha}{2} = ED$.

Arc $BC = r\alpha$, $C_{\text{shaded}} = 2r\sin\dfrac{\alpha}{2} + 2r + r\alpha$.

$S_{\triangle ABC} = \dfrac{1}{2}r^2 \sin\alpha$, $S_{\text{sector } ABC} = \dfrac{1}{2}r^2 \alpha$.

$S_{\text{shaded}} = 2r^2 \sin\dfrac{\alpha}{2} - \left(\dfrac{1}{2}r^2\alpha - \dfrac{1}{2}r^2\sin\alpha\right)$.

Example 例 <4.17>

The picture below shows two circles with centers A and C (Fig. 4.24). Circle C has radius 13 cm. Circle A has 12 cm. Two circles intersect at D and B where DB is the diameter of circle A. Also, CA is perpendicular to DB (Fig. 4.24).

(a) Figure out angle DCA in radians.

(b) Figure out the area of shaded region.

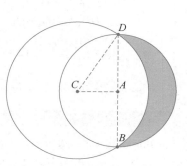

Fig. 4.24

Solution 解

(a) By Pythagoras in right triangle CAD, we have $CD = 13$, $AD = 12$, $CA = 5$, and angle $DCA = \sin^{-1}\dfrac{12}{13} = 1.176 \approx 1.18$.

(b) $S_{\triangle DCB} = 2S_{\triangle DCA} = 12 \times 5 = 60$.

$S_{\text{sector } DCB} = \dfrac{1}{2} \times 13 \times 13 \times 2 \times 1.176\,0 = 198.7$.

$S_{\text{semicircle } DB} = \dfrac{1}{2}\pi r^2 = 1/2 \times 12 \times 12 \times \pi = 72\pi$.

$S_{\text{shaded}} = S_{\text{semicircle } DB} - (S_{\text{sector } DCB} - S_{\triangle DCB}) = 87.449\,7 \approx 87.4$.

Example 例 <4.18>

From the Fig. 4.25, AED and ABC are radii of a circle with center A. AED and ABC are straight lines and $AD = 1.5r$ cm and $AE = r$ cm. The perimeter of the shaded region $BCDE$ is $6r$ (cm)^2.

(a) Find the value of angle DAC. Correct to four decimals places.

(b) When the area of the shaded region is 45 square centimeters. Figure out the value of r.

Solution 解

(a) Let $\angle EAB = \theta \Rightarrow r\theta + 1.5r\theta + 2(1.5r - r) = 6r \Rightarrow 2.5r\theta + r = 6r \Rightarrow \theta = 2.$

(b) $\frac{1}{2}(1.5r)^2 \cdot 2 - \frac{1}{2} \cdot r^2 \cdot 2 = 45 \Rightarrow r^2 = 36 \Rightarrow r = 6.$

Example 例 4.19

$EBDC$ is a plate which made by two parts. BDC is a semicircle. CEB is a segment of a circle with center A and radius r cm. Angle $BAC = \alpha$ (Fig. 4.26)

(a) Express the radius of the semicircle in r and α.

(b) Express the perimeter of the plate and area of the plate in r and α.

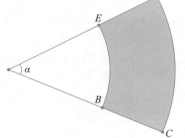

Fig. 4.25

Solution 解

(a) Major Arc $CEB = r \times (2\pi - \alpha)$.

$BC = 2r\sin\frac{\alpha}{2}.$

$r_{\text{semicircle}} = \frac{1}{2}BC = r\sin\frac{\alpha}{2}.$

(b) $C_{\text{plate}} = r \times (2\pi - \alpha) + \pi \times r\sin\frac{\alpha}{2}.$

$S_{\text{plate}} = S_{\text{semicircle}} + S_{\text{major arc}} + S_{\triangle ABC}$

$= \frac{1}{2}\pi\left(r\sin\frac{\alpha}{2}\right)^2 + \pi r^2 - \frac{1}{2}\alpha r^2 + \frac{1}{2}r^2.$

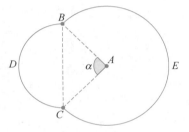

Fig. 4.26

Example 例 4.20

CD is a major arc of a circle A and radius 10 cm. Points G and F lie on AD and AC respectively. CD is tangent at E to the arc FEG which is a smaller circle also with center A. Angle $DAC = 2$ radians (Fig. 4.27).

(a) Figure out the radius of the arc FEG, correct to 4 decimal places.

(b) Find the area of the shaded region.

Solution 解

(a) $\cos 1 = \frac{AE}{AC} = \frac{AE}{10} \Rightarrow AE = 10 \times \cos 1 = 5.4030.$

(b) $S_{\text{large sector}} = \frac{1}{2} \times 10^2 \times (2\pi - 2) = 214.1593.$

$S_{\text{small sector}} = \frac{1}{2} \times 5.4030^2 \times 2 = 29.1924.$

$S_{\text{shaded}} = 214.159 + 29.1924 = 243.3517 \approx 243.$

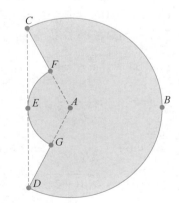

Fig. 4.27

Example 例 4.21

In the picture, BAC is a quarter circle with center A and radius 10 cm. The point D lies on the arc BC and

AEC is a straight line. The line DE is parallel to BA and angle $DAB = \pi/6$ (Fig. 4.28).

(a) Figure out the perimeter of the shaded region.
(b) Figure out the area of the shaded region.

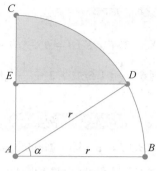

Fig. 4.28

Solution 解

(a) Angle EDA = Angle $DAB = \dfrac{\pi}{6}$.

$\dfrac{EA}{AD} = \sin \dfrac{\pi}{6} \Rightarrow EA = 10 \times \dfrac{1}{2} = 5 \Rightarrow EA = EC = 5$.

$ED = \sqrt{10^2 - 5^2} = 5\sqrt{3}$, Length of arc $CB = \dfrac{1}{2}\pi r = 5\pi$.

Length of arc $BD = r\alpha = \dfrac{5\pi}{3}$.

$C_{\text{shaded}} = 5 + 5\sqrt{3} + 5\pi + \dfrac{5\pi}{3}$.

(b) $S_{\text{quarter}} = \dfrac{1}{4}\pi r^2 = 25\pi$.

$S_{\triangle AED} = 5 \times 5\sqrt{3} \times \dfrac{1}{2} = \dfrac{25\sqrt{3}}{2}$.

$S_{\text{sector}} = \dfrac{1}{2}r^2 \times \alpha = \dfrac{25\pi}{3}$.

$S_{\text{shaded}} = S_{\text{quarter}} - S_{\triangle AED} - S_{\text{sector}} = \dfrac{50\pi}{3} - \dfrac{25\sqrt{3}}{2}$.

Example 例 4.22

Here is a right triangle. AB is perpendicular to CA. ADB is a straight line with $AD = DB = x$ cm. Angle $DCB = \alpha$ and angle $ABC = \beta$ radians (Fig. 4.29).

(a) Figure out CD in terms of x and β.
(b) Show that $\alpha = \left(\dfrac{\pi}{2} - \beta\right) - \tan^{-1}\left(\dfrac{1}{2\tan \beta}\right)$.

Solution 解

(a) $\tan \beta = \dfrac{CA}{AB} = \dfrac{CA}{2x} \Rightarrow CA = 2x \times \tan \beta$.

$CD = \sqrt{x^2 + (2x\tan \beta)^2}$.

(b) $\tan DCA = \dfrac{AD}{AC} = \dfrac{x}{2x\tan \beta} = \dfrac{1}{2\tan \beta}$.

$\alpha = \left(\dfrac{\pi}{2} - \beta\right) - \tan^{-1}\left(\dfrac{1}{2\tan \beta}\right)$.

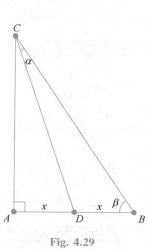

Fig. 4.29

Example 4.23

There are three circles with center C, B, and A. E, F, and D are the only three points which one circle touch with other two. $a = 15$ cm, $b = 14$ cm and $c = 6$ cm. what is the area of the shaded region (Fig. 4.30).

Solution

By Pythagoras, we can find that ABC is a right triangle where AC is perpendicular to CB at point C.

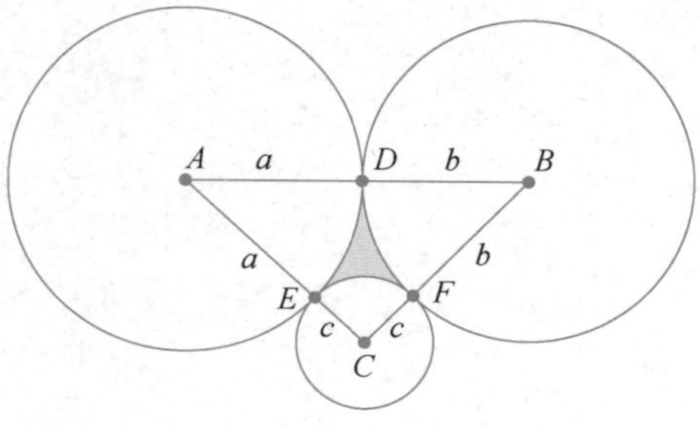

Let angle CAB be α and angle CBA be β.

$\alpha = \sin^{-1}\dfrac{20}{29} = \cos^{-1}\dfrac{21}{29}$.

$\beta = \sin^{-1}\dfrac{21}{29} = \cos^{-1}\dfrac{20}{29}$.

$S_{\triangle abc} = 20 \times 21 \times \dfrac{1}{2} = 210$.

$S_{\text{shaded}} = S_{\triangle abc} - \text{threesectors} = 210 - \dfrac{1}{2} \times \left(15 \times 15 \times \sin^{-1}\dfrac{20}{29} + 14 \times 14 \times \sin^{-1}\dfrac{21}{29} + 6 \times 6 \times \dfrac{\pi}{2}\right) = 16.8$.

Fig. 4.30

Summary of Key Theories 核心定义总结

(1) If the length of an arc is equal to the radius of the circle subtended at an angle θ, then the angle θ is said to have a measure of 1 radian.

(2) The arc length l of a sector with angle θ radians is determined by $l = r\theta$, where r is the radius of the circle.

(3) The area A of a sector with angle θ radians is determined by $A = \dfrac{1}{2}r^2\theta$, where r is the radius of the circle.

(4) Sine Law: $\dfrac{\sin A}{a} = \dfrac{\sin B}{b} = \dfrac{\sin C}{c}$.

(5) Cosine Law: $a^2 = b^2 + c^2 - 2bc\cos A$, or
$b^2 = a^2 + c^2 - 2ac\cos B$, or
$c^2 = a^2 + b^2 - 2ab\cos C$.

(6) Given triangle ABC and the sides are named to correspond to the opposite angles, the area of the triangle would be $\dfrac{1}{2}ab\sin C$.

(7) Given triangle ABC with corresponding sides a, b and c respectively. The area of the triangle would be $\sqrt{s(s-a)(s-b)(s-c)}$ where $s = \dfrac{a+b+c}{2}$.

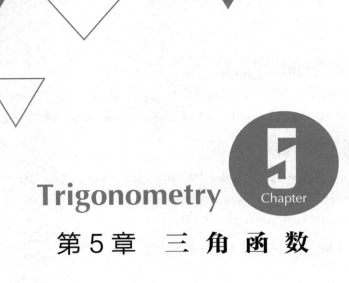

Trigonometry

第 5 章 三 角 函 数

We first encounter $\sin x$, $\cos x$ and $\tan x$ as ratios of sides in a right-angled triangle having x as one of the acute angles. These ratios depend only on the angle x, not on the particular triangle. We already had some reviews in last chapter, therefore, this chapter would begin with some definitions of angles.

> **Definition** 定义
>
> An acute angle is an angle less than 90 degree.
> A right angle is an angle equal to 90 degree.
> An obtuse angle in an angle between 90 and 180 degrees.
> A straight angle is an angle equal to 180 degree.
> A reflex angle in an angle between 180 and 360 degrees.
> A full angle is an angle equal to 360 degree.
>
> 锐角是小于 90° 的角.
> 直角是等于 90° 的角.
> 钝角是大于 90° 小于 180° 的角.
> 平角是等于 180° 的角.
> 反射角是大于 180° 小于 360° 的角.
> 全角是等于 360° 的角.

We should be familiar with the trigonometric functions when the angles are acute. However, we need more general definitions of trigonometric functions defined for all real numbers not just acute angles. Such definitions are phrased in terms of a circle rather than a triangle.

> **Definition** 定义
>
> A unit circle is the circle centered at the origin and radius 1, its standard equation is $x^2 + y^2 = 1$.
> 一个单位圆是以原点为圆心且半径为 1 的圆, 它的标准方程为 $x^2 + y^2 = 1$.

Consider a unit circle C, let $A(1,0)$ be a point on the circle C, and let P_t be the point on C at a distance $|t|$ measured along the circle from point A for any real number t.

Hence, P_t has coordinate $(\cos t, \sin t)$ because the radius is 1.

If t is positive, we measured along C in the counterclockwise direction.

If t is negative, we measured along C in the clockwise direction.

We are simply using the arc-length t as a measure of the angle AOP_t in radians under this construction.

Cosine and Sine are often called circular functions because they are defined this way (Fig. 5.1 and Fig. 5.2).

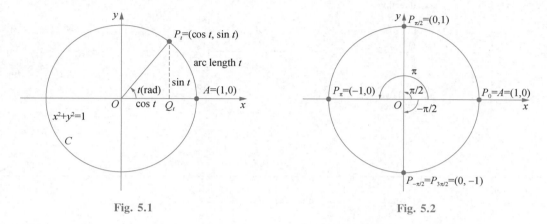

Fig. 5.1　　　　　　　　　　Fig. 5.2

There are some exact trigonometric ratios for certain angles which can be found from special triangles (Fig. 5.3).

$$\sin\left(\frac{\pi}{6}\right) = \frac{1}{2}, \cos\left(\frac{\pi}{6}\right) = \frac{\sqrt{3}}{2}, \tan\left(\frac{\pi}{6}\right) = \frac{1}{\sqrt{3}}.$$

$$\sin\left(\frac{\pi}{4}\right) = \frac{1}{\sqrt{2}}, \cos\left(\frac{\pi}{4}\right) = \frac{1}{\sqrt{2}}, \tan\left(\frac{\pi}{4}\right) = 1.$$

$$\sin\left(\frac{\pi}{3}\right) = \frac{\sqrt{3}}{2}, \cos\left(\frac{\pi}{3}\right) = \frac{1}{2}, \tan\left(\frac{\pi}{3}\right) = \sqrt{3}.$$

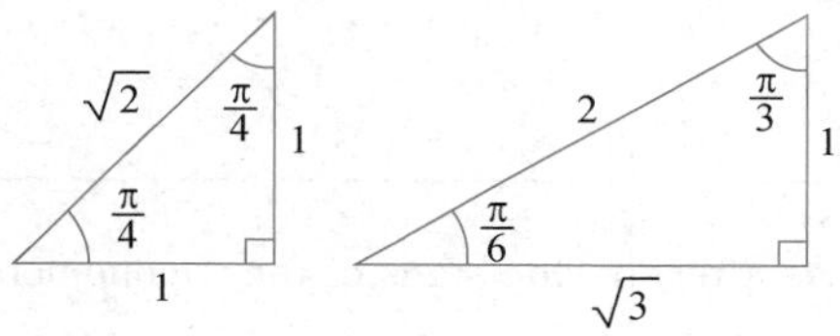

Fig. 5.3

We can also use these circular definitions to prove $\sin^2 t + \cos^2 t = 1$.

Since P_t has x-coordinate and y-coordinate, $\sin t$ lies on a unit circle for all real numbers t (angles measured in radian), therefore, $-1 \leq \sin t \leq 1$ and $-1 \leq \cos t \leq 1$.

The unit circle has equation $x^2 + y^2 = 1 \Rightarrow \cos^2 t + \sin^2 t = 1$.

While $\tan t = \dfrac{\sin t}{\cos t}$, so $\tan t$ does not exist when $\cos t = 0$; this implies that $\tan t$ is not defined when $t = (2n+1)\dfrac{\pi}{2}$ for n being integers, and $-\infty < \tan t < +\infty$.

5.1　Angle Measurement　角的测量

Suppose P lies on the unit circle, let θ be the angle measured from the positive x-axis.

If θ is positive, we use counter-clockwise direction; if θ is negative, we use clockwise direction (Fig. 5.4). For example, according to the diagram below, when we have $\theta = 210°$ and $\phi = -150°$, they represented the same angle (Fig. 5.5).

We can have an investigation for the signs of the trigonometric ratios in the four quadrants. We discovered and summarized in the following table (Tab. 5.1 and Fig. 5.6).

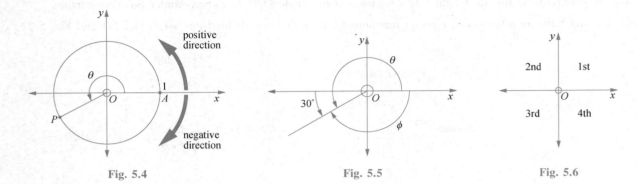

Fig. 5.4　　　　　　　Fig. 5.5　　　　　　　Fig. 5.6

Tab. 5.1

Quadrant	$\sin\theta$	$\cos\theta$	$\tan\theta$
1	+	+	+
2	+	−	−
3	−	−	+
4	−	+	−

As a result, the signs of the trigonometric ratios in each of the four quadrants can be remembered by means of the rule "All Students Take Chinese" as shown in the following figure (Fig. 5.7).

Note 注意

When using a calculator to calculate any trigonometric ratios, make sure you have selected the proper angular mode: Degrees or Radians.

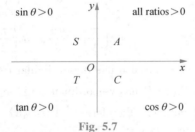

Fig. 5.7

5.2　Cosine is an Even Function and Sine is an Odd Function　余弦是偶函数，正弦是奇函数

Since the circle $x^2 + y^2 = 1$ is symmetric about x-axis, the points P_{-t} and P_t have the same x-coordinates but opposite y-coordinates (Fig. 5.8). Hence, $\cos(-t) = \cos t$ and $\sin(-t) = -\sin t$.

5.3 Complementary Angle Identities 余角等式

Two angles are complementary if their sum is $\frac{\pi}{2}$ radians. The points $P_{\frac{\pi}{2}-t}$ and P_t are the reflections of each other in the line $y = x$, so the x-coordinate of $P_{\frac{\pi}{2}-t}$ is the y-coordinate of P_t and vice versa (Fig. 5.9).

$$\cos\left(\frac{\pi}{2} - t\right) = \sin t, \quad \sin\left(\frac{\pi}{2} - t\right) = \cos t.$$

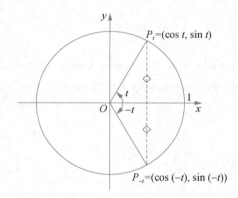

Fig. 5.8

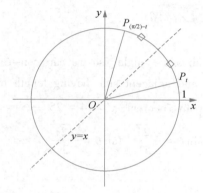

Fig. 5.9

5.4 Supplementary Angle Identities 补角等式

Two angles are supplementary if their sum is π radians. Since the circle $x^2 + y^2 = 1$ is symmetric about y-axis, the points $P_{\pi-t}$ and P_t have the same y-coordinates but opposite x-coordinates (Fig. 5.10). Hence,
$\cos(\pi - t) = -\cos t$ and $\sin(\pi - t) = \sin t$.

As a result, we can use these rules to find the trigonometric ratios of positive or negative angles using the corresponding acute angle θ made with the positive x-axis (Fig. 5.11).

$$\sin(180° - \theta°) = \sin\theta°, \qquad \sin(\pi - \theta) = \sin\theta.$$
$$\sin(180° + \theta°) = -\sin\theta°, \quad \text{or} \quad \sin(\pi + \theta) = -\sin\theta.$$
$$\sin(360° - \theta°) = -\sin\theta°, \qquad \sin(2\pi - \theta) = -\sin\theta.$$

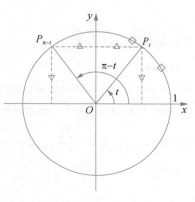

Fig. 5.10

$$\cos(180° - \theta°) = -\cos\theta°, \qquad \cos(\pi - \theta) = -\cos\theta.$$
$$\cos(180° + \theta°) = -\cos\theta°, \quad \text{or} \quad \cos(\pi + \theta) = -\cos\theta.$$
$$\cos(360° - \theta°) = \cos\theta°, \qquad \cos(2\pi - \theta) = \cos\theta.$$

$$\tan(180° - \theta°) = -\tan\theta°, \qquad \tan(\pi - \theta) = -\tan\theta.$$
$$\tan(180° + \theta°) = \tan\theta°, \quad \text{or} \quad \tan(\pi + \theta) = \tan\theta.$$
$$\tan(360° - \theta°) = -\tan\theta°, \qquad \tan(2\pi - \theta) = -\tan\theta.$$

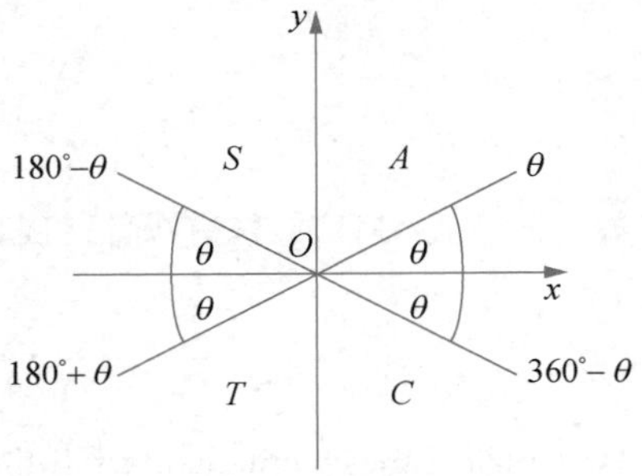

Fig. 5.11

Example 例 5.1

Given $\cos\theta = \dfrac{2}{5}$, and θ is an acute angle. Find the values of $\sin\theta$ and $\tan\theta$ (Fig. 5.12)

Solution 解

Method 1

Since it is an acute angle, so we can construct a right triangle with hypotenuse having length of 5 and adjacent side having length of 2 and opposite side having length of x. By Pythagorean Theorem, $x^2 + 4 = 25 \Rightarrow x = \sqrt{21}$.

Therefore, $\sin\theta = \dfrac{\sqrt{21}}{5}$, $\tan\theta = \dfrac{\sqrt{21}}{2}$.

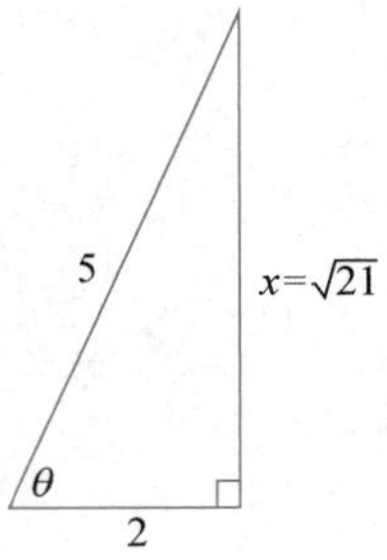

Fig. 5.12

Method 2

Using the trigonometric identity: $\sin^2\theta + \cos^2\theta = 1 \Rightarrow \sin^2\theta + \left(\dfrac{2}{5}\right)^2 = 1 \Rightarrow \sin\theta = \pm\dfrac{\sqrt{21}}{5}$.

Since it is acute angle; so it is in quadrant one.

Therefore, $\sin\theta = \dfrac{\sqrt{21}}{5}$, $\tan\theta = \dfrac{\sin\theta}{\cos\theta} = \dfrac{\sqrt{21}}{2}$.

Example 例 5.2

Given $\cos\theta = -\dfrac{2}{5}$, and θ is an obtuse angle. Find the values of $\sin\theta$ and $\tan\theta$.

Solution 解

Method 1

Even though it is an obtuse angle, we can still treat it as an acute angle, so we can construct a right triangle as shown (In quadrant 2, $\cos\theta$ is negative; but in quadrant 1 it is positive.).

Therefore, $\sin\theta = \dfrac{\sqrt{21}}{5}$, $\tan\theta = -\dfrac{\sqrt{21}}{2}$.

Method 2

Using the trigonometric identity: $\sin^2\theta + \cos^2\theta = 1 \Rightarrow \sin^2\theta + \left(\dfrac{2}{5}\right)^2 = 1 \Rightarrow \sin\theta = \pm\dfrac{\sqrt{21}}{5}$.

Since it is obtuse angle, it is in quadrant two.

Therefore, $\sin\theta = \dfrac{\sqrt{21}}{5}$, $\tan\theta = \dfrac{\sin\theta}{\cos\theta} = -\dfrac{\sqrt{21}}{2}$.

Example 例 5.3

Given $\sin\theta = -\dfrac{\sqrt{21}}{5}$, and θ is a reflex angle. Find the values of $\cos\theta$ and $\tan\theta$.

Solution 解

Method 1

Similar to example 5.2, we still treat it as an acute angle; so we can construct a right triangle as shown.
It is a reflex angle, so it might be in quadrant 3 or 4 ($\sin\theta$ is negative in both quadrant 3 and 4).

If it is in quadrant 3: $\cos\theta = \dfrac{-2}{5}$, $\tan\theta = \dfrac{\sqrt{21}}{2}$.

If it is in quadrant 4: $\cos\theta = \dfrac{2}{5}$, $\tan\theta = -\dfrac{\sqrt{21}}{2}$.

Method 2

Using the trigonometric identity: $\sin^2\theta + \cos^2\theta = 1 \Rightarrow \cos\theta = \pm\dfrac{2}{5}$.

If it is in quadrant 3: $\cos\theta = \dfrac{-2}{5}$, $\tan\theta = \dfrac{\sqrt{21}}{2}$.

If it is in quadrant 4: $\cos\theta = \dfrac{2}{5}$, $\tan\theta = -\dfrac{\sqrt{21}}{2}$.

Example 例 5.4

Given $\sin 34° = a$, find the following in terms of a.
(a) $\cos 34°$; (b) $\sin 214°$; (c) $\sin 146°$; (d) $\cos 146°$; (e) $\cos 214°$; (f) $\sin 326°$; (g) $\cos 326°$.

Solution 解

$34°$ is in quadrant one.
(a) $\cos 34° = \sqrt{1-a^2}$.
(b) $\sin 214° = -\sin 34° = -a$.
(c) $\sin 146° = \sin(180° - 34°) = \sin 34° = a$.
(d) $\cos 146° = \cos(180° - 34°) = -\cos 34° = -\sqrt{1-a^2}$.
(e) $\cos 214° = \cos(180° + 34°) = -\cos 34° = -\sqrt{1-a^2}$.
(f) $\sin 326° = \sin(360° + 34°) = -\sin 34° = -a$.
(g) $\cos 326° = \cos(360° + 34°) = \cos 34° = \sqrt{1-a^2}$.

Example 例 5.5

Given $\sin\theta + \cos\theta = \dfrac{1}{\sqrt{2}}$. Find the values of the followings:

(a) $\sin\theta\cos\theta$; (b) $\tan\theta + \cot\theta$; (c) $\sin\theta - \cos\theta$.

Solution 解

(a) When we square both sides of the given condition, we would obtain $\sin^2\theta + 2\sin\theta\cos\theta + \cos^2\theta = \frac{1}{2} \Rightarrow$
$$1 + 2\sin\theta\cos\theta = \frac{1}{2} \Rightarrow \sin\theta\cos\theta = \frac{-1}{4}.$$

(b) $\tan\theta + \cot\theta = \frac{\sin\theta}{\cos\theta} + \frac{\cos\theta}{\sin\theta} = \frac{\sin^2\theta + \cos^2\theta}{\sin\theta\cos\theta} = \frac{1}{-1/4} = -4.$

(c) $(\sin\theta - \cos\theta)^2 = (\sin\theta + \cos\theta)^2 - 4\sin\theta\cos\theta = \frac{1}{2} - 4 \times \frac{-1}{4} = \frac{3}{2}.$

Hence, $\sin\theta - \cos\theta = \pm\sqrt{\frac{3}{2}}.$

Example 例 5.6

Given $\tan\theta + \cot\theta = 3$, and θ is an acute angle. Find the values of the followings:
(a) $\sin\theta\cos\theta$; (b) $\sin\theta + \cos\theta$; (c) $\sin\theta - \cos\theta$.

Solution 解

(a) $\tan\theta + \cot\theta = \frac{\sin\theta}{\cos\theta} + \frac{\cos\theta}{\sin\theta} = \frac{\sin^2\theta + \cos^2\theta}{\sin\theta\cos\theta} = \frac{1}{\sin\theta\cos\theta} = 3 \Rightarrow \sin\theta\cos\theta = \frac{1}{3}.$

(b) $(\sin\theta + \cos\theta)^2 = 1 + 2\sin\theta\cos\theta = \frac{5}{3}.$ Since it is an acute angle, so both $\sin\theta$ and $\cos\theta$ are positive.

Hence, $\sin\theta + \cos\theta = \sqrt{\frac{5}{3}}.$

(c) $(\sin\theta - \cos\theta)^2 = 1 - 2\sin\theta\cos\theta = \frac{1}{3} \Rightarrow \sin\theta - \cos\theta = \pm\sqrt{\frac{1}{3}}.$

Example 例 5.7

Given $\sin\theta + \cos\theta = \frac{7}{5}$. Find the values of $\sin\theta$ and $\cos\theta$.

Solution 解

Method 1

$\sin\theta + \cos\theta = \frac{7}{5} \Rightarrow \sin\theta = \frac{7}{5} - \cos\theta$; squaring both sides $\Rightarrow \sin^2\theta = \frac{49}{25} - \frac{14}{5}\cos\theta + \cos^2\theta$

$\Rightarrow 1 - \cos^2\theta = \frac{49}{25} - \frac{14}{5}\cos\theta + \cos^2\theta \Rightarrow 25\cos^2\theta - 35\cos\theta + 12 = 0$

$\Rightarrow (5\cos\theta - 3)(5\cos\theta - 4) = 0 \Rightarrow \cos\theta = \frac{3}{5} \text{ or } \frac{4}{5}.$

When $\cos\theta = \frac{3}{5} \Rightarrow \sin\theta = \frac{4}{5}.$

When $\cos\theta = \frac{4}{5} \Rightarrow \sin\theta = \frac{3}{5}.$

Method 2

$(\sin\theta + \cos\theta)^2 = 1 + 2\sin\theta\cos\theta = \dfrac{49}{25} \Rightarrow \sin\theta\cos\theta = \dfrac{12}{25}.$

$(\sin\theta - \cos\theta)^2 = 1 - 2\sin\theta\cos\theta = \dfrac{1}{25} \Rightarrow \sin\theta - \cos\theta = \pm\dfrac{1}{5}.$

When $\sin\theta - \cos\theta = \dfrac{1}{5} \Rightarrow \sin\theta = \dfrac{4}{5}$, $\cos\theta = \dfrac{3}{5}$.

When $\sin\theta - \cos\theta = -\dfrac{1}{5} \Rightarrow \sin\theta = \dfrac{3}{5}$, $\cos\theta = \dfrac{4}{5}$.

Method 3

Consider a quadratic equation with roots $\sin\theta$ and $\cos\theta$.

Since $\sin\theta + \cos\theta = \dfrac{7}{5} \Rightarrow \sin\theta\cos\theta = \dfrac{12}{25}.$

Therefore, the quadratic equation can be written as $x^2 - \dfrac{7}{5}x + \dfrac{12}{25} = 0 \Rightarrow 25x^2 - 35x + 12 = 0 \Rightarrow$
$(5x - 3)(5x - 4) = 0 \Rightarrow x = \dfrac{3}{5}, \dfrac{4}{5}.$

When $\sin\theta = \dfrac{4}{5} \Rightarrow \cos\theta = \dfrac{3}{5}$.

When $\sin\theta = \dfrac{3}{5} \Rightarrow \cos\theta = \dfrac{4}{5}$. ∎

5.5 Trigonometric Identities 三角恒等式

A trigonometric identity is a relationship among the trigonometric functions which is always true for all values of the angle θ.

三角恒等式是三角函数之间的关系，对于所有的角 θ 都成立.

The followings are the basic trigonometric identities we have already known:

$$\tan\theta = \dfrac{\sin\theta}{\cos\theta}, \quad \csc\theta = \dfrac{1}{\sin\theta}, \quad \sec\theta = \dfrac{1}{\cos\theta},$$
$$\cot\theta = \dfrac{1}{\tan\theta}, \quad \sin^2\theta + \cos^2\theta = 1.$$

From the identity $\sin^2\theta + \cos^2\theta = 1$.

When we divided the identity by $\cos^2\theta$, we obtained $\tan^2\theta + 1 = \sec^2\theta$.

When we divided the identity by $\cos^2\theta$, we obtained $1 + \cot^2\theta = \csc^2\theta$.

In this section, we will use the known identities to prove the more complicated identities.

The basic steps for proving trigonometric identities are as follows:
(1) Begin with the more complicated side.
(2) Change all expressions to $\sin\theta$ and $\cos\theta$.
(3) Use some algebraic manipulation to complete the proof.

Example 例 5.8

Prove the identity $\sec\theta - \dfrac{\cos\theta}{1+\sin\theta} = \tan\theta$.

Solution 解

$$\text{LHS} = \sec\theta - \frac{\cos\theta}{1+\sin\theta} = \frac{1}{\cos\theta} - \frac{\cos\theta}{1+\sin\theta} = \frac{1+\sin\theta-\cos^2\theta}{\cos\theta(1+\sin\theta)} = \frac{1+\sin\theta-(1-\sin^2\theta)}{\cos\theta(1+\sin\theta)}$$

$$= \frac{1+\sin\theta-1+\sin^2\theta}{\cos\theta(1+\sin\theta)} = \frac{\sin\theta+\sin^2\theta}{\cos(1+\sin\theta)} = \frac{\sin\theta(1+\sin\theta)}{\cos\theta(1+\sin\theta)} = \frac{\sin\theta}{\cos\theta} = \tan\theta = \text{RHS}.$$

Note 注意

LHS is the abbreviation for left-hand side, and RHS is the abbreviation for right-hand side.

Example 例 5.9

Prove the identity $1 - \tan^2\theta = \dfrac{\cos^4\theta - \sin^4\theta}{\cos^2\theta}$.

Solution 解

$$\text{RHS} = \frac{\cos^4\theta - \sin^4\theta}{\cos^2\theta} = \frac{(\cos^2\theta-\sin^2\theta)(\cos^2\theta+\sin^2\theta)}{\cos^2\theta} = \frac{(\cos^2\theta-\sin^2\theta)\times 1}{\cos^2\theta}$$

$$= \frac{\cos^2\theta}{\cos^2\theta} - \frac{\sin^2\theta}{\cos^2\theta} = 1 - \tan^2\theta = \text{LHS}.$$

Example 例 5.10

Prove the identity $\csc\theta - 1 = \dfrac{\cos\theta}{\tan\theta(1+\sin\theta)}$.

Solution 解

$$\text{RHS} = \frac{\cos\theta}{\tan\theta(1+\sin\theta)} = \frac{\cos\theta}{\dfrac{\sin\theta}{\cos\theta}(1+\sin\theta)} = \frac{\cos^2\theta}{\sin\theta(1+\sin\theta)} = \frac{1-\sin^2\theta}{\sin\theta(1+\sin\theta)}$$

$$= \frac{(1+\sin\theta)(1-\sin\theta)}{\sin\theta(1+\sin\theta)} = \frac{(1-\sin\theta)}{\sin\theta} = \frac{1}{\sin\theta} - 1 = \csc\theta - 1 = \text{LHS}.$$

Example 例 5.11

Prove the identity $(\cot\theta + \csc\theta)^2 = \dfrac{1+\cos\theta}{1-\cos\theta}$.

Solution 解

$$\text{LHS} = (\cot\theta + \csc\theta)^2 = \left(\frac{\cos\theta}{\sin\theta} + \frac{1}{\sin\theta}\right)^2 = \left(\frac{\cos\theta + 1}{\sin\theta}\right)^2 = \frac{(\cos\theta + 1)^2}{\sin^2\theta}$$

$$= \frac{(\cos\theta + 1)^2}{1 - \cos^2\theta} = \frac{(\cos\theta + 1)^2}{(1 - \cos\theta)(1 + \cos\theta)} = \frac{(\cos\theta + 1)}{(1 - \cos\theta)} = \text{RHS}.$$

5.6 Graph of Trigonometric Functions 三角函数的图像

The unit circle has circumference 2π which means adding 2π to t causes the point P_t to go one extra complete revolution around the unit circle and end up in the same place; that is, $P_{t+2\pi} = P_t$. Hence, the sine and cosine functions are periodic functions.

Definition 定义

$f(x)$ is a periodic function with period p if and only if $f(x + p) = f(x)$, $\forall x$; and p is the smallest positive value for the relationship to be hold.

$f(x)$ 是周期为 p 的周期函数，当且当 $f(x + p) = f(x)$, $\forall x$; 并且 p 是关系成立的最小正值.

Consider a wave oscillates about a horizontal line which is called the principal axis as shown in Fig. 5.13.

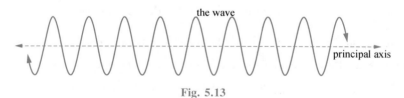

Fig. 5.13

The maximum point occurs at the top of a crest, and the minimum point occurs at the bottom of a trough. The amplitude is the distance between a maximum and the principal axis or the distance between a minimum point and the principal axis (Fig. 5.14).

The mathematical equations for the principal axis is: $y = \dfrac{\text{max. value} + \text{min. value}}{2}$.

The amplitude A: $A = \dfrac{\text{max. value} - \text{min. value}}{2}$.

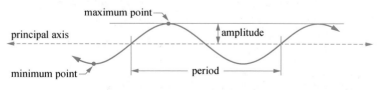

Fig. 5.14

5.7 The Sine Curve 正弦曲线

We already know a point P lies on a unit circle can be described by $(\cos x, \sin x)$ where x is the angle makes with the positive x-axis. The values of y-coordinate of P made up the sine curve, so if we project the values of y-coordinate of P to a set of axes alongside, we can obtain the basic sine curve.

Beware that we should assume that angle x is measured in radians unless it is indicated that the angles are in degrees. The range is $-1 \leq y \leq 1$, domain is **R**. The period of basic sine curve is 2π (Fig. 5.15).

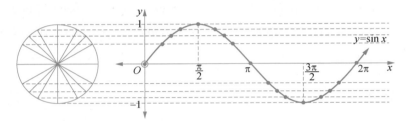

Fig. 5.15

We can use the knowledge we learned in Chapter 2 (transformation of functions) to graph the general sine function $y = a\sin[b(x-c)] + d$ assuming a, b, c and d are positive; then the graph of $y = a\sin[b(x-c)] + d$ is a horizontal translation of c units to the right followed by a horizontal stretched of a factor of $\dfrac{1}{b}$; and a vertical stretched by a factor of a followed by a vertical translation of d units upwards. Since the sine curve is a periodic curve, so it is handy to draw the curve by finding its min., max., period and principal axis.

The graph of $y = a\sin[b(x-c)] + d$, assuming a and b are positive (for negative values, we treated them as positive and then do a reflection to obtain the final result).

This graph has period of $\dfrac{2\pi}{|b|}$, the principal axis is $y = d$. The max. value is $a + d$, and the min. value is $-a + d$. The amplitude of the curve is $|a|$.

After finding all of the required values, we first divided the period by 4; since it is a sine curve so we begin with the point (c, d) and then move to the max. point with coordinate $\left(c + \dfrac{2\pi}{4b}, a + d\right)$. After reaching the max. point, the curve travel back to the principal axis with coordinate $\left(c + 2 \times \dfrac{2\pi}{4b}, d\right)$, and then move to the min. point with coordinate $\left(c + 3 \times \dfrac{2\pi}{4b}, -a + d\right)$. Finally, it comes back to the principal axis with coordinate $\left(c + 4 \times \dfrac{2\pi}{4b}, d\right)$ which completes a period of the graph.

Example 5.12

Sketch the graph of $y = 2\sin\left(3x - \dfrac{3\pi}{4}\right) + 1$.

Solution

We first rewrite the equation $y = 2\sin\left(3x - \dfrac{3\pi}{4}\right) + 1 = 2\sin\left(3\left(x - \dfrac{\pi}{4}\right)\right) + 1$.

Amplitude: 2, Period: $\dfrac{2\pi}{3}$, Max.: $2 + 1 = 3$, Min.: $-2 + 1 = -1$, Principal axis: $y = 1$.

It is a sine curve (Fig. 5.16), so the graph start with $\left(\dfrac{\pi}{4}, 1\right)$ moves to the max. $\left(\dfrac{\pi}{4} + \dfrac{2\pi}{12}, 3\right)$, then back to principal axis $\left(\dfrac{\pi}{4} + 2 \times \dfrac{2\pi}{12}, 1\right)$; and it moves to the min. $\left(\dfrac{\pi}{4} + 3 \times \dfrac{2\pi}{12}, -1\right)$, then travel back to the principal axis $\left(\dfrac{\pi}{4} + 4 \times \dfrac{2\pi}{12}, 1\right)$.

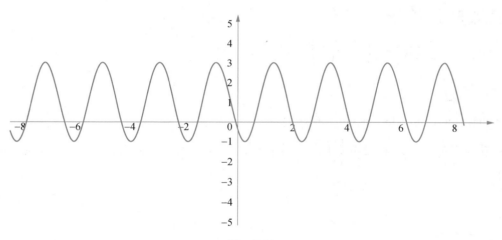

Fig. 5.16

Example 5.13

Sketch the graph of $y = -2\sin\left[3\left(x - \dfrac{\pi}{4}\right)\right] + 1$.

Solution

Amplitude: $|-2| = 2$, Period: $\dfrac{2\pi}{3}$, Max.: $2 + 1 = 3$, Min.: $-2 + 1 = -1$, Principal axis: $y = 1$.

It is a sine curve (Fig. 5.17), so the graph start with $\left(\dfrac{\pi}{4}, 1\right)$ moves to the min. $\left(\dfrac{\pi}{4} + \dfrac{2\pi}{12}, -1\right)$, then back to principal axis $\left(\dfrac{\pi}{4} + 2 \times \dfrac{2\pi}{12}, 1\right)$; and it moves to the max. $\left(\dfrac{\pi}{4} + 3 \times \dfrac{2\pi}{12}, 3\right)$, then travel back to the principal axis $\left(\dfrac{\pi}{4} + 4 \times \dfrac{2\pi}{12}, 1\right)$. It is basically a reflection about the principal axis from the graph of $y = 2\sin\left[3\left(x - \dfrac{\pi}{4}\right)\right] + 1$.

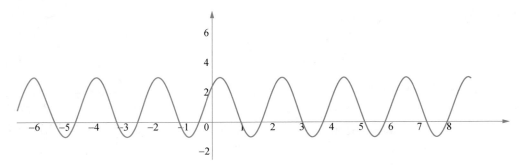

Fig. 5.17

Example 例 5.14

Sketch the graph of $y = 2\sin\left[-3\left(x - \dfrac{\pi}{4}\right)\right] + 1$.

Solution 解

Use the fact that sine is an odd function, $f(-x) = -f(x)$.

$y = 2\sin\left[-3\left(x - \dfrac{\pi}{4}\right)\right] + 1 = -2\sin\left[3\left(x - \dfrac{\pi}{4}\right)\right] + 1$, so the graph would be the same as example 5.13.

5.8 The Cosine Curve 余弦曲线

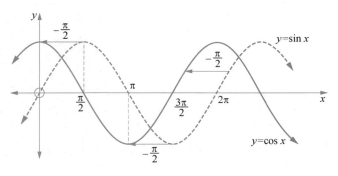

Fig. 5.18

We already know that $\sin\left(\dfrac{\pi}{2} - x\right) = \cos x$; so any cosine curve is just a horizontal shift of the sine curve. The basic cosine curve is shown in the Fig. 5.18. The range is $-1 \leqslant y \leqslant 1$, domain is $\mathbf{R}$. The period of basic cosine curve is 2π.

The graph of $y = a\sin[b(x - c)] + d$, assuming a and b are positive (for negative values, we treated them as positive and then do a reflection to obtain the final result).

This graph has period of $\frac{2\pi}{|b|}$, the principal axis is $y = d$. The max. value is $a + d$, and the min. value is $-a + d$. The amplitude of the curve is $|a|$.

After finding all of the required values, we first divided the period by 4; since it is a cosine curve, we begin at the max. point with coordinate $(c, a + d)$, then move to the principal axis with coordinate $\left(c + \frac{2\pi}{4b}, d\right)$, and then move to the min. point with coordinate $\left(c + 2 \times \frac{2\pi}{4b}, -a + d\right)$; then come back to the principal axis with coordinate $\left(c + 3 \times \frac{2\pi}{4b}, d\right)$. Finally, it moves to the max. point with coordinate $\left(c + 4 \times \frac{2\pi}{4b}, a + d\right)$ which completes a period of the graph.

Example 5.15

Sketch the graph of $y = 3\cos[\pi(x + 1)] - 2$.

Solution

Amplitude: 3, Period: $\frac{2\pi}{\pi} = 2$, Max.: $3 - 2 = 1$, Min.: $-3 - 2 = -5$, Principal axis: $y = -2$.

It is a cosine curve (Fig. 5.19), so the graph start with the max. point $(-1, 1)$ and then moves to the principal axis $\left(-1 + \frac{2}{4}, -2\right)$, then go to the min. point $\left(-1 + 2 \times \frac{2}{4}, -5\right)$; and travel back to the principal axis $\left(-1 + 3 \times \frac{2}{4}, -2\right)$. Finally, it goes to the max. point $\left(-1 + 4 \times \frac{2}{4}, 1\right)$ which completes a period of the graph.

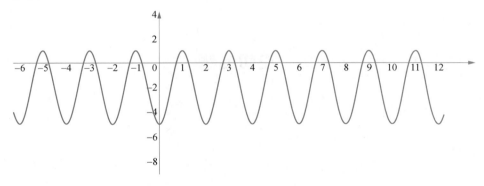

Fig. 5.19

Example 5.16

Sketch the graph of $y = -3\cos[\pi(x + 1)] - 2$.

Solution

Amplitude: $|-3| = 3$, Period: $\frac{2\pi}{\pi} = 2$, Max.: $3 - 2 = 1$, Min.: $-3 - 2 = -5$, Principal axis: $y = -2$.

It is a cosine curve (Fig. 5.20), so the graph starts at the min. point $(-1, -5)$ and then moves to the principal axis $\left(-1+\dfrac{2}{4},-2\right)$, then go to the max. point $\left(-1+2\times\dfrac{2}{4},1\right)$; and travel back to the principal axis $\left(-1+3\times\dfrac{2}{4},-2\right)$. Finally, it goes to the min. point $\left(-1+4\times\dfrac{2}{4},-5\right)$ which completes a period of the graph. ∎

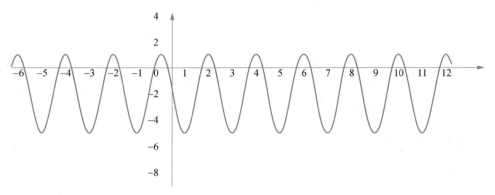

Fig. 5.20

Example 例 5.17

Sketch the graph of $y=3\cos[-\pi(x+1)]-2$.

Solution 解

Use the fact that cosine is an even function, $f(-x)=f(x)$.
$y=3\cos[-\pi(x+1)]-2=3\cos[\pi(x+1)]-2$, so the graph would be the same as example 5.16. ∎

5.9 The Tangent Curve 正切曲线

Since $\tan x=\dfrac{\sin x}{\cos x}$, so it is undefined when $\cos x=0$; that means the graph of $y=\tan x$ has vertical asymptotes when $x=\pm\dfrac{(2n-1)\pi}{2}$, where $n\in\mathbf{Z}^+$.
The basic tangent curve is shown in Fig. 5.21.
The range is $\mathbf{R}$, domain is $x\neq\dfrac{(2n-1)\pi}{2}$, where $n\in\mathbf{Z}^+$. The period of basic tangent curve is π.

Note 注意

There is no amplitude for tangent curve.

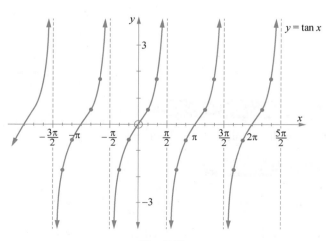

Fig. 5.21

The graph of $y = a\tan[b(x-c)] + d$, assuming a and b are positive (for negative values, we treated them as positive and then do a reflection to obtain the final result).

This graph has period of $\dfrac{\pi}{|b|}$, the vertical asymptote occurs at $x = c \pm \dfrac{(2n-1)\pi}{2b}$, where $n \in \mathbf{Z}^+$ and the principal axis is $y = d$.

正切曲线没有振幅.如 $y = a\tan[b(x-c)] + d$ 的图像，假设 a 和 b 是正值（对于负值，可以先把它们当作正值，然后做一个反射就能得到最终的结果），该图的周期为 $\dfrac{\pi}{|b|}$，纵渐近线出现在 $x = c \pm \dfrac{(2n-1)\pi}{2b}$，其中 $n \in \mathbf{Z}^+$，主轴是 $y = d$.

Example 例 5.18

Sketch the graph of $y = \tan\left[\dfrac{\pi}{3}(x-2)\right] + 1$.

Solution 解

Amplitude：None，Period：$\dfrac{\pi}{\pi/3} = 3$，Principal axis：$y = 1$，Vertical Asymptotes：$x = 2 \pm \dfrac{3(2n-1)}{2}$，$n \in \mathbf{Z}^+$.

It is a tangent curve (Fig. 5.22), so the graph passing through the principal axis at $(2, 1)$.

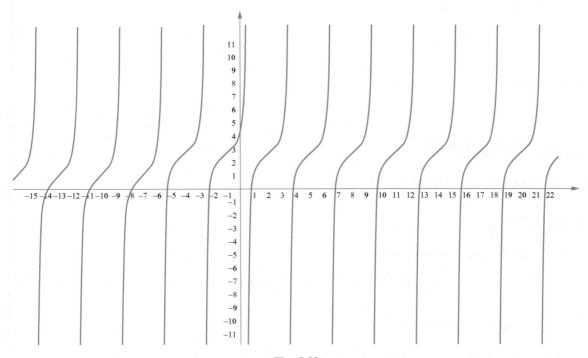

Fig. 5.22

Example 例 5.19

Sketch the graph of $y = -\tan\left[\dfrac{\pi}{3}(x-2)\right] + 1$.

Solution 解

Amplitude: None, Period: $\dfrac{\pi}{\pi/3} = 3$, Principal axis: $y = 1$, Vertical Asymptotes: $x = 2 \pm \dfrac{3(2n-1)}{2}$, $n \in \mathbf{Z}^+$.

It is a tangent curve (Fig. 5.23), so the graph passing through the principal axis at $(2, 1)$; and it is a reflection about the principal axis of the graph of $y = \tan\left[\dfrac{\pi}{3}(x-2)\right] + 1$.

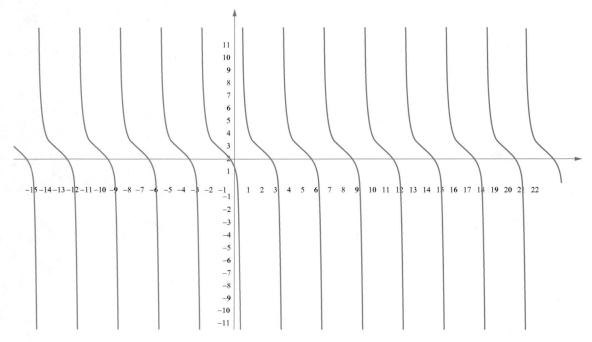

Fig. 5.23

Example 例 <5.20>

Sketch the graph of $y = \tan\left[-\dfrac{\pi}{3}(x-2)\right] + 1$.

Solution 解

Use the fact that tangent is an odd function, $f(-x) = -f(x)$.

$y = \tan\left[-\dfrac{\pi}{3}(x-2)\right] + 1 = -\tan\left[\dfrac{\pi}{3}(x-2)\right] + 1$, so the graph would be the same as example 5.19.

Example 例 <5.21>

The diagram shows parts of the graph of the sine function. Please write out the equation of the function.

Solution 解

Max: 4, Min: 2, Principal axis: $y = 3$, Horizontal shift to the left by 2 units (Fig. 5.24).
Amplitude: 1, Period: 4.
So the equation can be written as $y = a\sin[b(x+c)] + d$.

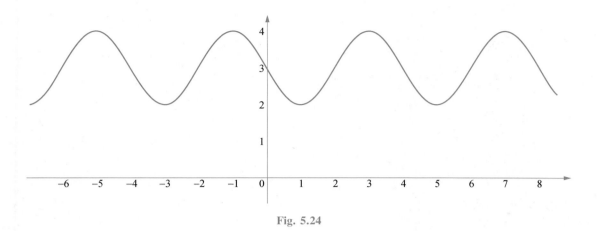

Fig. 5.24

Period = $\dfrac{2\pi}{b} \Rightarrow 4 = \dfrac{2\pi}{b} \Rightarrow b = \dfrac{\pi}{2}$, d = principal axis = 3, $a + 3$ = max. = 4 $\Rightarrow a = 1$.

$c = -2$. Hence, $y = \sin\left[\dfrac{\pi}{2}(x + 2)\right] + 3$ is one of the possible solutions. ∎

Note 注意

We can also say this graph is shift to the right by 2 units, then the equation became $y = \sin\left[\dfrac{\pi}{2}(x - 2)\right] + 3$, or we can write it as $y = \cos\left[\dfrac{\pi}{2}(x - 3)\right] + 3$.

5.10 Inverse Trigonometric Functions 反三角函数

The six trigonometric functions are periodic (we did not discuss the graphs of cosec x, sec x and cot x). Hence, they are not one-to-one functions. However, as we did in Chapter 2, we can restrict their domain in such a way that they are one-to-one, therefore they are invertible.

这 6 个三角函数是周期函数(这里先不讨论 csc x,sec x 和 cot x 的图像),因此,它们不是一对一函数.然而,正如在第 2 章中讲解过的,可以通过限制它们的定义域使它们成为一对一函数,因此,它们是可逆的.

5.11 The Inverse Sine (or Arcsine) Function 反正弦函数

Let $f(x) = \sin x$ for $\dfrac{-\pi}{2} \leqslant x \leqslant \dfrac{\pi}{2}$ (Fig. 5.25), then it is a one-to-one function. Being one-to-one, $f(x) = \sin x$ has an inverse which is denoted $\sin^{-1} x$ or $\arcsin x$ and it is called the inverse sine or arcsine function.

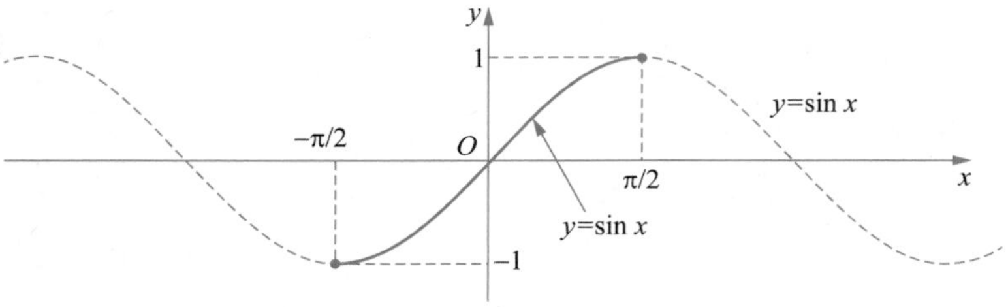

Fig. 5.25

📖 Definition 定义

$y = \sin^{-1} x \Leftrightarrow x = \sin y$ and $\dfrac{-\pi}{2} \leqslant y \leqslant \dfrac{\pi}{2}$.

Cancellation identities:

$\sin^{-1}(\sin x) = \arcsin(\sin x) = x$ if $\dfrac{-\pi}{2} \leqslant x \leqslant \dfrac{\pi}{2}$.

$\sin(\sin^{-1} x) = \sin(\arcsin x) = x$ if $-1 \leqslant x \leqslant 1$.

The graph of $y = \sin^{-1} x$ is the reflection of the graph of $y = \sin x$ with domain $-\dfrac{\pi}{2} \leqslant x \leqslant \dfrac{\pi}{2}$ in the line $y = x$. The domain of $y = \sin^{-1} x$ is $[-1, 1]$ and the range is $\left[\dfrac{-\pi}{2}, \dfrac{\pi}{2}\right]$ (Fig. 5.26).

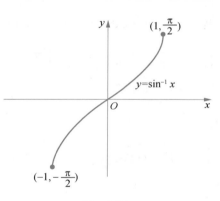

Fig. 5.26

✏️ Example 例 <5.22>

Find the values of the followings, the angles are measured in radians.

(a) $\sin^{-1}\left(\dfrac{-1}{\sqrt{2}}\right)$;   (b) $\sin^{-1}(-1)$;

(c) $\sin^{-1}\left(\dfrac{1}{2}\right)$;   (d) $\sin^{-1}\left(\dfrac{3}{2}\right)$.

✍️ Solution 解

(a) $\sin^{-1}\left(\dfrac{-1}{\sqrt{2}}\right) = \dfrac{-\pi}{4}$ because $\sin\left(-\dfrac{\pi}{4}\right) = -\dfrac{1}{\sqrt{2}}$ and $\dfrac{-\pi}{2} \leqslant \dfrac{-\pi}{4} \leqslant \dfrac{\pi}{2}$.

(b) $\sin^{-1}(-1) = \dfrac{-\pi}{2}$ because $\sin\left(-\dfrac{\pi}{2}\right) = -1$ and $\dfrac{-\pi}{2} \leqslant \dfrac{-\pi}{2} \leqslant \dfrac{\pi}{2}$.

(c) $\sin^{-1}\left(\dfrac{1}{2}\right) = \dfrac{\pi}{6}$ because $\sin\left(\dfrac{\pi}{6}\right) = \dfrac{1}{2}$ and $\dfrac{-\pi}{2} \leqslant \dfrac{\pi}{6} \leqslant \dfrac{\pi}{2}$.

(d) $\sin^{-1}\left(\dfrac{3}{2}\right)$ is not defined because $\dfrac{3}{2}$ is not in the range of $y = \sin x$.

Example 5.23

Find the values of the followings, the angles are measured in radians.

(a) $\sin(\sin^{-1} 0.5)$; (b) $\sin^{-1}(\sin 0.7)$; (c) $\sin^{-1}\left(\sin \dfrac{5\pi}{6}\right)$; (d) $\cos(\sin^{-1} 0.6)$.

Solution 解

(a) $\sin(\sin^{-1} 0.5) = 0.5$ by cancellation identities.

(b) $\sin^{-1}(\sin 0.7) = 0.7$ by cancellation identities.

(c) $\sin^{-1}\left(\sin \dfrac{5\pi}{6}\right) \neq \dfrac{5\pi}{6}$ because $\dfrac{5\pi}{6}$ is not in the interval $\left[-\dfrac{\pi}{2}, \dfrac{\pi}{2}\right]$; so we can not apply cancellation identities. We know $\sin \dfrac{5\pi}{6} = \sin \dfrac{\pi}{6}$, and $\dfrac{\pi}{6}$ is in the interval $\left[-\dfrac{\pi}{2}, \dfrac{\pi}{2}\right]$. Therefore,

$$\sin^{-1}\left(\sin \dfrac{5\pi}{6}\right) = \sin^{-1}\left(\sin \dfrac{\pi}{6}\right) = \dfrac{\pi}{6}.$$

(d) Let $\theta = \sin^{-1} 0.6 \Rightarrow \sin \theta = 0.6 = \dfrac{3}{5}$, so $\cos(\sin^{-1} 0.6) = \cos \theta = \sqrt{1 - \sin^2 \theta} = \dfrac{4}{5}$.

Example 5.24

Find the expressions of $\tan(\sin^{-1} x)$ and $\cos(\sin^{-1} x)$.

Solution 解

Let $\theta = \sin^{-1} x \Rightarrow \sin \theta = x = \dfrac{x}{1}$, so we can construct a right triangle as shown in Fig. 5.27.

$\tan(\sin^{-1} x) = \tan \theta = \dfrac{x}{\sqrt{1-x^2}}$ and $\cos(\sin^{-1} x) = \cos \theta = \sqrt{1-x^2}$.

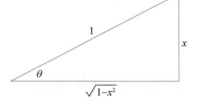

Fig. 5.27

5.12 The Inverse Tangent (or Arctangent) Function
反正切函数

The inverse tangent (or arctangent) function is defined similar to the arcsine function. Let $f(x) = \tan x$ for $\dfrac{-\pi}{2} < x < \dfrac{\pi}{2}$, then it is a one-to-one function. Being one-to-one, $f(x) = \tan x$ has an inverse which is denoted $\tan^{-1} x$ or $\arctan x$ and it is called the inverse tangent or arctangent function (Fig. 5.28).

反正切函数的定义类似于反正弦函数.假设$f(x)=\tan x$, $\dfrac{-\pi}{2} < x < \dfrac{\pi}{2}$, 那么,它是一个一对一函数.因为是一对一函数,所以, $f(x)=\tan x$ 有一个逆函数 $\tan^{-1} x$, 称为反正切函数(图 5.28).

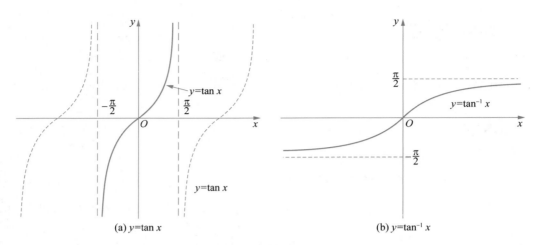

Fig. 5.28

🔱 Definition   定义

$y = \tan^{-1} x \Leftrightarrow x = \tan y$ and $\dfrac{-\pi}{2} < y < \dfrac{\pi}{2}$.

Cancellation identities:

$\tan^{-1}(\tan x) = \arctan(\tan x) = x$ if $\dfrac{-\pi}{2} < x < \dfrac{\pi}{2}$.

$\tan(\tan^{-1} x) = \tan(\arctan x) = x$ if $-\infty < x < \infty$.

📖 Example 例 ⟨5.25⟩

Find the values of the followings, the angles are measured in radians.

(a) $\tan(\tan^{-1} -5)$;   (b) $\tan^{-1}\left(\tan \dfrac{\pi}{4}\right)$;   (c) $\tan^{-1}\left(\tan \dfrac{7\pi}{4}\right)$;   (d) $\cos(\tan^{-1} 3)$.

✍ Solution 解

(a) $\tan(\tan^{-1} -5) = -5$ by cancellation identities.

(b) $\tan^{-1}\left(\tan \dfrac{\pi}{4}\right) = \dfrac{\pi}{4}$ by cancellation identities.

(c) $\tan^{-1}\left(\tan \dfrac{7\pi}{4}\right) \neq \dfrac{7\pi}{4}$ because $\dfrac{7\pi}{4}$ is not in the interval $\left[-\dfrac{\pi}{2}, \dfrac{\pi}{2}\right]$; so we can not apply cancellation identities. We know $\tan \dfrac{7\pi}{4} = \tan \dfrac{3\pi}{4} = \tan \dfrac{-\pi}{4}$, and $-\dfrac{\pi}{4}$ is in the interval $\left[-\dfrac{\pi}{2}, \dfrac{\pi}{2}\right]$ but $\dfrac{3\pi}{4}$. Therefore, $\tan^{-1}\left(\tan \dfrac{7\pi}{4}\right) = \tan^{-1}\left(\tan \dfrac{-\pi}{4}\right) = \dfrac{-\pi}{4}$.

(d) Let $\theta = \tan^{-1} 3 \Rightarrow \tan \theta = 3 = \dfrac{3}{1}$, so we can construct a right angled triangle with hypotenuse of length $\sqrt{10}$, opposite side of length 3 and adjacent side of length 1. So $\cos(\tan^{-1} 3) = \cos \theta = \dfrac{1}{\sqrt{10}}$.

Example 例 5.26

Find the expressions of $\sin(\tan^{-1} x)$ and $\cos(\tan^{-1} x)$.

Solution 解

Let $\tan^{-1} x = \theta \Rightarrow \tan \theta = x = \dfrac{x}{1}$, so we can construct a right-angled triangle with opposite side of length x and adjacent side with length 1. By Pythagoras' Theorem, the hypotenuse would have length of $\sqrt{1 + x^2}$.

Therefore, $\sin(\tan^{-1} x) = \sin \theta = \dfrac{x}{\sqrt{1 + x^2}}$ and $\cos(\tan^{-1} x) = \cos \theta = \dfrac{1}{\sqrt{1 + x^2}}$. ■

5.13 The Inverse Cosine (or Arccosine) Function
反余弦函数

The function $f(x) = \cos x$ is one-to-one on the interval $[0, \pi]$, so the inverse $\cos^{-1} x$ or $\arccos x$ exists. If $y = \cos^{-1} x \Leftrightarrow \cos y = x$ and $0 \leq y \leq \pi$. Since we know $\cos y = \sin\left(\dfrac{\pi}{2} - y\right)$, so $-\dfrac{\pi}{2} \leq \dfrac{\pi}{2} - y \leq \dfrac{\pi}{2} \Rightarrow 0 \leq y \leq \pi$. Thus, we would obtain $y = \cos^{-1} x \Leftrightarrow x = \sin\left(\dfrac{\pi}{2} - y\right) \Leftrightarrow \sin^{-1} x = \dfrac{\pi}{2} - y = \dfrac{\pi}{2} - \cos^{-1} x$ (Fig. 5.29).

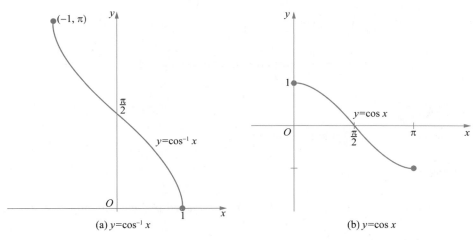

(a) $y = \cos^{-1} x$ (b) $y = \cos x$

Fig. 5.29

函数 $f(x) = \cos x$ 在区间 $[0, \pi]$ 上是一对一函数,所以,它的反函数 $\cos^{-1} x$ 是存在的.假设 $y = \cos^{-1} x \Leftrightarrow \cos y = x$,且 $0 \leq y \leq \pi$.我们知道 $\cos y = \sin\left(\dfrac{\pi}{2} - y\right)$,那么, $-\dfrac{\pi}{2} \leq \dfrac{\pi}{2} - y \leq \dfrac{\pi}{2} \Rightarrow 0 \leq y \leq \pi$.因此,可以推导出 $y = \cos^{-1} x \Leftrightarrow x = \sin\left(\dfrac{\pi}{2} - y\right) \Leftrightarrow \sin^{-1} x = \dfrac{\pi}{2} - y = \dfrac{\pi}{2} - \cos^{-1} x$ (图 5.29).

Definition 定义

$y = \cos^{-1} x \Leftrightarrow x = \cos y$ and $0 \leqslant y \leqslant \pi$.

$\cos^{-1} x = \dfrac{\pi}{2} - \sin^{-1} x$ for $-1 \leqslant x \leqslant 1$.

Cancellation identities：

$\cos^{-1}(\cos x) = \arccos(\cos x) = x$ if $0 \leqslant x \leqslant \pi$.

$\cos(\cos^{-1} x) = \cos(\arccos x) = x$ if $-1 \leqslant x \leqslant 1$.

Example 例 5.27

Find the values of the followings, the angles are measured in radians.

(a) $\cos(\cos^{-1} 0.4)$；(b) $\cos^{-1}\left(\cos\dfrac{\pi}{4}\right)$；(c) $\cos^{-1}\left(\cos\dfrac{5\pi}{4}\right)$；(d) $\sin(\cos^{-1} 0.5)$.

Solution 解

(a) $\cos(\cos^{-1} 0.4) = 0.4$ by cancellation identities.

(b) $\cos^{-1}\left(\cos\dfrac{\pi}{4}\right) = \dfrac{\pi}{4}$ by cancellation identities.

(c) $\cos^{-1}\left(\cos\dfrac{5\pi}{4}\right) \neq \dfrac{5\pi}{4}$ because $\dfrac{5\pi}{4}$ is not in the interval $[0, \pi]$, so we can not apply cancellation identities. We know $\cos\dfrac{5\pi}{4} = \cos\dfrac{3\pi}{4}$, and $\dfrac{3\pi}{4}$ is in the interval $[0, \pi]$.

Therefore, $\cos^{-1}\left(\cos\dfrac{5\pi}{4}\right) = \cos^{-1}\left(\cos\dfrac{3\pi}{4}\right) = \dfrac{3\pi}{4}$.

(d) Let $\theta = \cos^{-1} 0.5 \Rightarrow \cos\theta = 0.5 = \dfrac{1}{2}$, $\sin(\cos^{-1} 0.5) = \sin\theta = \sqrt{1 - \cos^2\theta} = \dfrac{\sqrt{3}}{2}$.

Example 例 5.28

Find the expressions of $\sin(\cos^{-1} x)$ and $\tan(\cos^{-1} x)$.

Solution 解

Let $\cos^{-1} x = \theta \Rightarrow \cos\theta = x = \dfrac{x}{1}$, so we can construct a right-angled triangle with adjacent side of length x and hypotenuse of length 1. By Pythagoras' Theorem, the opposite side would have length of $\sqrt{1 - x^2}$.

Therefore, $\sin(\cos^{-1} x) = \sin\theta = \sqrt{1 - x^2}$ and $\tan(\cos^{-1} x) = \tan\theta = \dfrac{\sqrt{1 - x^2}}{x}$.

Chapter 5　Trigonometry

5.14　Solving Trigonometric Equations　解三角方程

We have to keep in mind that in order to solve a trigonometric equation, we must convert the trigonometry ratios to the same one. According to the general curves of basic trigonometric functions, we know there are multiple solutions for a trigonometric equation if they existed. Generally speaking, for $\sin x$ and $\cos x$, we usually have two solutions in one period; while for $\tan x$, we usually have one solution in one period if they existed.

我们必须记住,为了解一个三角方程,必须把三角比转换成相同的.根据基础三角函数的一般曲线,可以知道一个三角方程如果有解,可以存在多个解.一般来说,对于$\sin x$和$\cos x$,通常在一个周期内存在两个解;对于$\tan x$,如果有解的话,通常在一个周期内存在一个解.

Example 例 5.29

Solve $3\sin x = -1$, for $0° \leqslant x \leqslant 360°$.

Solution 解

$3\sin x = -1 \Rightarrow \sin x = \dfrac{-1}{3} \Rightarrow x = -19.471\,22°$. Since this angle is not in the specified domain, so we have to find the corresponding angles.

$-19.471\,22° \equiv 340.528\,78°$ which is in quadrant 4; and another angle would be in quadrant 3 which is equal to $180° + 19.471\,22° = 199.471\,22°$.

So $x_1 = 340.5°$ and $x_2 = 199.5°$ corrected to one decimal places.

Example 例 5.30

Solve $3\sin x = 1$, for $0 \leqslant x \leqslant 2\pi$.

Solution 解

$3\sin x = 1 \Rightarrow \sin x = \dfrac{1}{3} \Rightarrow x = 0.339\,836\,9$ which is in the specified domain, so another angle would be in quadrant 2 which is equal to $\pi - 0.339\,836\,9 = 2.801\,756$.

So $x_1 = 0.340$ and $x_2 = 2.80$ corrected to three significant figures.

Example 例 5.31

Solve $\sin x = 1$, for $0 \leqslant x \leqslant 2\pi$.

Solution 解

$\sin x = 1 \Rightarrow x = \dfrac{\pi}{2}$, and 1 it is the maximum value for $\sin x$; hence in one period we only have one solution.

So $x = \dfrac{\pi}{2}$ only.

Example 5.32

Solve $\sin x = -1$, for $0 \leqslant x \leqslant 2\pi$.

Solution

$\sin x = -1 \Rightarrow x = \dfrac{3\pi}{2}$, and -1 it is the minimum value for $\sin x$; hence in one period we only have one solution. So $x = \dfrac{3\pi}{2}$ only. ∎

Example 5.33

Solve $\sin x = 0$, for $0 \leqslant x \leqslant 2\pi$.

Solution

$\sin x = 0 \Rightarrow x = 0$, π and 2π. The principal axis for the curve $y = \sin x$ is $y = 0$, so in this case we have three solutions. ∎

Note 注意

For trigonometric equation expressed in terms of $\sin x$, the equation has two solutions in one period if it is not the maximum nor minimum nor the value of principal axis. It has one solution in one period if it is equal to maximum or minimum values. It has three solutions in one period when the equation is equal to the value of the principal axis but we will have five solutions in two periods, seven solutions in three periods and so on.

对于用 $\sin x$ 表示的三角方程，如果它不是最大值、不是最小值，也不是主轴值，该方程在一个周期内有两个解；如果它等于最大值或最小值，该方程在一个周期内有一个解；当方程等于主轴的值，该方程在一个周期内有 3 个解，但在两个周期内有 5 个解，在 3 个周期内有 7 个解，依此类推.

Example 5.34

Solve $3\sin 2x = 1$, for $0 \leqslant x \leqslant 2\pi$.

Solution

$3\sin 2x = 1 \Rightarrow \sin 2x = \dfrac{1}{3}$, since $\sin 2x$ has a period of $\dfrac{2\pi}{2} = \pi$ with maximum value of 1 and minimum value of -1; therefore in this interval $[0, 2\pi]$ we should expect 4 solutions.

$\sin 2x = \dfrac{1}{3} \Rightarrow 2x = 0.339\,84 \Rightarrow x = 0.169\,92$ or $2x = \pi - 0.339\,84 = 2.801\,76 \Rightarrow x = 1.400\,88$.

Or $2x = 0.339\,84 + 2\pi \Rightarrow x = 0.169\,92 + \pi = 3.311\,51$ or $2x = 2.801\,76 + 2\pi \Rightarrow x = 1.400\,88 + \pi = 4.542\,47$.

So $x_1 = 0.170$, $x_2 = 1.40$, $x_3 = 3.31$ and $x_4 = 4.54$ corrected to three significant figures. ∎

Example 5.35

Solve $\sin 2x = 1$, for $0 \leqslant x \leqslant 2\pi$.

Solution

Since $\sin 2x$ has a period of $\dfrac{2\pi}{2} = \pi$ with maximum value of 1; therefore in this interval $[0, 2\pi]$ we should expect 2 solutions.

So $x_1 = \dfrac{\pi}{4}$ and $x_2 = \dfrac{5\pi}{4}$ are the solutions. ∎

Example 5.36

Solve $\sin 2x = 0$, for $0 \leqslant x \leqslant 2\pi$.

Solution

Since $\sin 2x$ has a period of $\dfrac{2\pi}{2} = \pi$ and 0 is the value of principal axis; therefore in this interval $[0, 2\pi]$ we should expect 5 solutions (not six solutions).

So $x_1 = 0$, $x_2 = \dfrac{\pi}{2}$, $x_3 = \pi$, $x_4 = \dfrac{3\pi}{2}$ and $x_5 = 2\pi$ are the solutions. ∎

Example 5.37

Solve $3\cos x = -1$, for $0° \leqslant x \leqslant 360°$.

Solution

$3\cos x = -1 \Rightarrow \cos x = \dfrac{-1}{3} \Rightarrow x = 109.47°$, the angle is in quadrant 2; so another angle would be in quadrant 3 which is equal to $360° - 109.47° = 250.53°$.

So $x_1 = 109.5°$ and $x_2 = 250.5°$ corrected to one decimal places. ∎

Example 5.38

Solve $3\cos x = 1$, for $0 \leqslant x \leqslant 2\pi$.

Solution

$3\cos x = 1 \Rightarrow \cos x = \dfrac{1}{3} \Rightarrow x = 1.230\,96$ which is in the quadrant 1, so another angle would be in quadrant 4 which is equal to $2\pi - 1.230\,96 = 5.052\,226$.

So $x_1 = 1.23$ and $x_2 = 5.05$ corrected to three significant figures. ∎

Example 5.39

Solve $\cos x = 1$, for $0 \leqslant x \leqslant 2\pi$.

Solution

$\cos x = 1 \Rightarrow x = 0$ and $x = 2\pi$. Even though 1 is the maximum value for $\cos x$; but in one period we still have two solutions which is different than the case for $\sin x$.

Example 例 ⟨5.40⟩

Solve $\cos x = -1$, for $0 \leqslant x \leqslant 2\pi$.

Solution 解

-1 it is the minimum value for $\cos x \Rightarrow x = \pi$, so in one period we only have one solution. So $x = \pi$ only.

Example 例 ⟨5.41⟩

Solve $\cos x = 0$, for $0 \leqslant x \leqslant 2\pi$.

Solution 解

$\cos x = 0 \Rightarrow x = \dfrac{\pi}{2}, \dfrac{3\pi}{2}$. Even though the principal axis for the curve $y = \cos x$ is $y = 0$, but but in one period we still have two solutions which is different than the case for $\sin x$.

Note 注意

For trigonometric equation expressed in terms of $\cos x$, the equation has two solutions in one period if it is not minimum value. It has one solution in one period if it is equal to the minimum values.

对于用 $\cos x$ 表示的三角方程,如果它不是最小值,该方程在一个周期内有两个解;如果它等于最小值,它在一个周期内有一个解.

Example 例 ⟨5.42⟩

Solve $3\cos 2x = 1$, for $0 \leqslant x \leqslant 2\pi$.

Solution 解

$3\cos 2x = 1 \Rightarrow \cos 2x = \dfrac{1}{3}$, since $\cos 2x$ has a period of $\dfrac{2\pi}{2} = \pi$ with maximum value of 1 and minimum value of -1; therefore in this interval $[0, 2\pi]$ we should expect 4 solutions.

$\cos 2x = \dfrac{1}{3} \Rightarrow 2x = 1.230\,96 \Rightarrow x = 0.615\,48$ or $2x = 2\pi - 1.230\,96 = 5.052\,22 \Rightarrow x = 2.526\,11$.

Or $2x = 1.230\,96 + 2\pi \Rightarrow x = 0.615\,48 + \pi = 3.757\,07$ or $2x = 5.052\,22 + 2\pi \Rightarrow x = 2.526\,11 + \pi = 5.667\,71$.

So $x_1 = 0.615$, $x_2 = 2.53$, $x_3 = 3.76$ and $x_4 = 5.67$ corrected to three significant figures.

Example 例 ⟨5.43⟩

Solve $3\tan x = -1$, for $0° \leqslant x \leqslant 360°$.

Solution 解

$3\tan x = -1 \Rightarrow \tan x = \dfrac{-1}{3} \Rightarrow x = -18.434\,95°$. Since this angle is not in the specified domain, so we have to find the corresponding angles.

$-18.43495° + 360° = 341.56505°$ which is in quadrant 4; and another angle would be in quadrant 2 which is equal to $180° - 18.43495° = 161.56505$.

So $x_1 = 341.6°$ and $x_2 = 161.6°$ corrected to one decimal places. ∎

Example 例 5.44

Solve $3\tan x = 1$, for $0 \leq x \leq 2\pi$.

Solution 解

$3\tan x = 1 \Rightarrow \tan x = \dfrac{1}{3} \Rightarrow x = 0.32175$ which is in the quadrant 1, so another angle would be in quadrant 3 which is equal to $\pi + 0.32175 = 3.46334$.

So $x_1 = 0.322$ and $x_2 = 3.46$ corrected to three significant figures. ∎

Note 注意

For trigonometric equation expressed in terms of $\tan x$, the equation has one solution in one period only.

对于用 $\tan x$ 表示的三角方程，该方程在一个周期内有一个解.

Example 例 5.45

Solve $3\tan 2x = 1$, for $0 \leq x \leq \pi$.

Solution 解

$3\tan 2x = 1 \Rightarrow \tan 2x = \dfrac{1}{3}$, since $\tan 2x$ has a period of $\dfrac{\pi}{2}$ therefore in this interval $[0, \pi]$ we should expect 2 solutions.

$\tan 2x = \dfrac{1}{3} \Rightarrow 2x = 0.32175 \Rightarrow x = 0.16088$ or $2x = \pi + 0.32175 \Rightarrow x = \dfrac{\pi}{2} + 0.16088 = 1.73167$.

So $x_1 = 0.161$ and $x_2 = 1.73$ corrected to three significant figures. ∎

Example 例 5.46

Solve the equation $5 - 7\sin x = 3\cos^2 x$ for $0 \leq x \leq 2\pi$.

Solution 解

The equation involves both $\sin x$ and $\cos x$, so we have to convert it into one single trigonometry ratio first. Use the identity $\sin^2 x + \cos^2 x = 1 \Rightarrow \cos^2 x = 1 - \sin^2 x$.

$5 - 7\sin x = 3\cos^2 x \Rightarrow 5 - 7\sin x = 3(1 - \sin^2 x) \Rightarrow 3\sin^2 x - 7\sin x + 2 = 0 \Rightarrow (\sin x - 2)(3\sin x - 1) = 0$

$\Rightarrow \sin x = 2$ or $\sin x = \dfrac{1}{3}$.

When $\sin x = 2$, there is no solution.

When $\sin x = \dfrac{1}{3} \Rightarrow x_1 = 0.33984$, $x_2 = \pi - 0.33984 = 2.80176$.

Therefore, $x_1 = 0.340$ and $x_2 = 2.80$ corrected to three significant figures. ∎

Example 例 <5.47>

Solve the equation $5\sin^2\theta - 2\sin\theta\cos\theta = 3\cos^2\theta$ for $0 \leqslant x \leqslant 2\pi$.

Solution 解

This example is different than example 5.45, if we use the identity $\sin^2 x + \cos^2 x = 1$ we still unable to get an equation involve only one trigonometry ratio because the term $2\sin\theta\cos\theta$. Therefore, we divided the equation by $\cos^2\theta$ to get $5\dfrac{\sin^2\theta}{\cos^2\theta} - 2\dfrac{\sin\theta}{\cos\theta} = 3 \Rightarrow 5\tan^2\theta - 2\tan\theta - 3 = 0 \Rightarrow (\tan\theta - 1)(5\tan\theta + 3) = 0 \Rightarrow \tan\theta = 1$ or $\tan\theta = \dfrac{-3}{5}$.

When $\tan\theta = 1$, $\theta_1 = \dfrac{\pi}{4}$ and $\theta_2 = \dfrac{5\pi}{4}$.

When $\tan\theta = \dfrac{-3}{5}$, $\theta_3 = -0.540\,42 \equiv 5.742\,77$ and $\theta_4 = 2.601\,17$.

So the solutions are $\theta_1 = \dfrac{\pi}{4}$, $\theta_2 = \dfrac{5\pi}{4}$, $\theta_3 = 5.74$ and $\theta_4 = 2.60$.

Example 例 <5.48>

Find the maximum and minimum values of $f(x) = 2\sin^2 x - 3\cos x + 5$.

Solution 解

$f(x) = 2\sin^2 x - 3\cos x + 5 = 2(1 - \cos^2 x) - 3\cos x + 5 = -2\cos^2 x - 3\cos x + 7$

$= -2\left(\cos^2 x + \dfrac{3}{2}\cos x\right) + 7 = -2\left(\cos^2 x + \dfrac{3}{2}\cos x + \dfrac{9}{16} - \dfrac{9}{16}\right) + 7$

$= -2\left(\cos^2 x + \dfrac{3}{2}\cos x + \dfrac{9}{16}\right) + 7 + \dfrac{9}{8} = -2\left(\cos x + \dfrac{3}{4}\right)^2 + \dfrac{65}{8}$.

So the maximum value is $\dfrac{65}{8}$ when $\cos x = \dfrac{-3}{4}$, the minimum value occurs when $\cos x = 1$ which is equal to 2.

Example 例 <5.49>

Find the maximum and minimum values of $f(x) = \dfrac{6\sin x + 9}{3\sin x + 4}$.

Solution 解

$f(x) = \dfrac{6\sin x + 9}{3\sin x + 4} = \dfrac{6\sin x + 8 + 1}{3\sin x + 4} = 2 + \dfrac{1}{3\sin x + 4}$.

Since $-1 \leqslant \sin x \leqslant 1$, so $f(x)$ is maximum when $\sin x = -1$; $f(x)$ is minimum when $\sin x = 1$.

Hence, the maximum value of $f(x)$ is 3 and the minimum value of $f(x)$ is $\dfrac{15}{7}$.

Summary of Key Theories　核心定义总结

(1) An acute angle is an angle less than 90 degree.

　　A right angle is an angle equal to 90 degree.

　　An obtuse angle in an angle between 90 and 180 degrees.

　　A straight angle is an angle equal to 180 degree.

　　A reflex angle in an angle between 180 and 360 degrees.

　　A full angle is an angle equal to 360 degree.

(2) A unit circle is the circle centered at the origin and radius 1, its standard equation is $x^2 + y^2 = 1$.

(3) $f(x)$ is a periodic function with period p if and only if $f(x+p) = f(x)$, $\forall x$; and p is the smallest positive value for the relationship to be hold.

(4) $y = \sin^{-1} x \Leftrightarrow x = \sin y$ and $\dfrac{-\pi}{2} \leqslant y \leqslant \dfrac{\pi}{2}$.

　　Cancellation identities:

　　$\sin^{-1}(\sin x) = \arcsin(\sin x) = x$ if $\dfrac{-\pi}{2} \leqslant x \leqslant \dfrac{\pi}{2}$.

　　$\sin(\sin^{-1} x) = \sin(\arcsin x) = x$ if $-1 \leqslant x \leqslant 1$.

(5) $y = \tan^{-1} x \Leftrightarrow x = \tan y$ and $\dfrac{-\pi}{2} < y < \dfrac{\pi}{2}$.

　　Cancellation identities:

　　$\tan^{-1}(\tan x) = \arctan(\tan x) = x$ if $\dfrac{-\pi}{2} < x < \dfrac{\pi}{2}$.

　　$\tan(\tan^{-1} x) = \tan(\arctan x) = x$ if $-\infty < x < +\infty$.

(6) $y = \cos^{-1} x \Leftrightarrow x = \cos y$ and $0 \leqslant y \leqslant \pi$.

　　$\cos^{-1} x = \dfrac{\pi}{2} - \sin^{-1} x$ for $-1 \leqslant x \leqslant 1$.

　　Cancellation identities:

　　$\cos^{-1}(\cos x) = \arccos(\cos x) = x$ if $0 \leqslant x \leqslant \pi$.

　　$\cos(\cos^{-1} x) = \cos(\arccos x) = x$ if $-1 \leqslant x \leqslant 1$.

Binomial Theorem

第 6 章 二项式定理

6.1 Pascal's Triangle 帕斯卡三角形

In algebra a sum of two terms such as $x + y$ is called a binomial. The binomial theorem is applying to expand a binomial expression that has been raised to some power.

We will only focus on positive integer n in this chapter.

Before get into the binomial theorem, we have to look at the Pascal's triangle first.

Consider the expansion for $(x + y)^n$ for first few positive integers n.

When $n = 1$, $x + y$.
When $n = 2$, $x^2 + 2xy + y^2$.
When $n = 3$, $x^3 + 3x^2y + 3xy^2 + y^3$.
When $n = 4$, $x^4 + 4x^3y + 6x^2y^2 + 4xy^3 + y^4$.

The coefficients in each terms forms a Pascal's triangle. The Pascal's triangle is formed by adding adjacent pairs of numbers to produce the number on next row (Fig. 6.1).

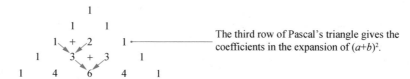

Fig. 6.1

When we investigate the terms in the expansion of $(x + y)^n$. We found out that there are $n + 1$ terms in the expansion, and the sum of exponents in each term is just equal to n.

Use Pascal's triangle to find the expansion of $(2x + 3y)^3$.

Chapter 6 Binomial Theorem

Solution 解

$(2x + 3y)^3 = (2x)^3 + 3(2x)^2(3y) + 3(2x)(3y)^2 + (3y)^3 = 8x^3 + 36x^2y + 54xy^2 + 27y^3$. ∎

Example 例 6.2

(a) Use Pascal's triangle to find the expansion of $(2x - 1)^4$;

(2) Find the coefficient of x^4 in the expansion of $(x + 3)(2x - 1)^4$.

Solution 解

(a) $(2x - 1)^4 = (2x)^4 + 4(2x)^3(-1) + 6(2x)^2(-1)^2 + 4(2x)(-1)^3 + (-1)^4 = 16x^4 - 32x^3 + 24x^2 - 8x + 1$.

(b) In order to get x^4, we need x^3 in the expansion of $(2x - 1)^4$ multiply by the term x in $(x + 3)$ plus the term of x^4 in the expansion of $(2x - 1)^4$ multiply by the term 3 in $(x + 3)$.

Therefore, the coefficient of x^4 is equal to $-32 + 3 \times 16 = 16$. ∎

We still need few more terminologies before introducing the binomial theorem. Let's begin with the factorial notation.

Definition 定义

For each positive integer n, the quantity n factorial denoted $n!$, is defined to be the product of all consecutive integers from 1 to n. $n! = (n - 1)!$, and $0! = 1$.

对于每个正整数 n，定义 n 的阶乘所表示的 $n!$ 为从 1 到 n 的所有连续整数的乘积. $n! = n(n-1)!$，并且 $0! = 1$.

The first few factorials are listed below：

$0! = 1$, $1! = 1$, $2! = 2 \times 1 = 2$, $3! = 3 \times 2 \times 1 = 6$, $4! = 4 \times 3 \times 2 \times 1 = 24$, $5! = 5 \times 4 \times 3 \times 2 \times 1 = 120$.

Example 例 6.3

Simplify the following expressions：

(a) $\dfrac{7!}{5!}$; (b) $\dfrac{6!}{3! \cdot 2!}$; (c) $\dfrac{1}{4! \cdot 2!} + \dfrac{1}{3! \cdot 2!}$; (d) $\dfrac{(n+1)!}{(n-1)!}$.

Solution 解

(a) $\dfrac{7!}{5!} = \dfrac{7 \times 6 \times 5!}{5!} = 42$.

(b) $\dfrac{6!}{3!2!} = \dfrac{6 \times 5 \times 4 \times 3!}{3! \cdot 2!} = 60$.

(c) $\dfrac{1}{4!2!} + \dfrac{1}{3!2!} = \dfrac{1}{4!2!} + \dfrac{1}{3!2!} \cdot \dfrac{4}{4} = \dfrac{1}{4!2!} + \dfrac{4}{4!2!} = \dfrac{5}{4!2!} = \dfrac{5}{48}$.

(d) $\dfrac{(n+1)!}{(n-1)!} = \dfrac{(n+1) \cdot n \cdot (n-1)!}{(n-1)!} = n(n+1) = n^2 + n$. ∎

👑 Definition 定义

The symbol $\binom{n}{r} = {}^nC_r$ which is read as "n choose r", denotes the total numbers of subset of size r that can be chosen from a set of n elements. $\binom{n}{r}$ are often called the binomial coefficients, because they arise in the binomial expansion.

$$\binom{n}{r} = {}^nC_r = \frac{n!}{(n-r)!\,r!}, \text{ for } 0 \leq r \leq n.$$

符号 $\binom{n}{r} = {}^nC_r$ 读作"n 选 r",表示 r 子集的总的数量可以从 n 的集合里选取. $\binom{n}{r}$ 常被称为二项式系数,是因为它们在二项式展开中出现.

$$\binom{n}{r} = {}^nC_r = \frac{n!}{(n-r)!\,r!}, \; 0 \leq r \leq n.$$

📖 Example 例 6.4

Compute the followings:

(a) $\binom{8}{5}$; (b) ${}^8C_3$; (c) $\binom{12}{7}$; (d) ${}^{12}C_5$.

✍ Solution 解

(a) $\binom{8}{5} = \dfrac{8!}{(8-5)!\,5!} = \dfrac{8 \times 7 \times 6 \times 5!}{3!\,5!} = 8 \times 7 = 56.$

(b) ${}^8C_3 = \dfrac{8!}{(8-3)!\,3!} = \dfrac{8 \times 7 \times 6 \times 5!}{5!\,3!} = 8 \times 7 = 56.$

(c) $\binom{12}{7} = \dfrac{12!}{(12-7)!\,7!} = \dfrac{12 \times 11 \times 10 \times 9 \times 8 \times 7!}{5!\,7!} = \dfrac{12 \times 11 \times 10 \times 9 \times 8}{5 \times 4 \times 3 \times 2 \times 1} = 792.$

(d) ${}^{12}C_5 = \dfrac{12!}{(12-5)!\,5!} = \dfrac{12 \times 11 \times 10 \times 9 \times 8 \times 7!}{7!\,5!} = \dfrac{12 \times 11 \times 10 \times 9 \times 8}{5 \times 4 \times 3 \times 2 \times 1} = 792.$

From the definition, it is clear that $\binom{n}{0} = \dfrac{n!}{n!\,0!} = 1$ and $\binom{n}{n} = \dfrac{n!}{0!\,n!} = 1.$

Also, $\binom{n}{1} = \dfrac{n!}{(n-1)!\,1!} = \dfrac{n \cdot (n-1)!}{(n-1)!\,1!} = n.$

In example 6.4, we can conclude a conjecture that $\binom{n}{r} = \binom{n}{n-r}$.

We will prove this statement in the next example.

📖 Example 例 6.5

Prove $\binom{n}{r} = \binom{n}{n-r}$.

Proof 证明

$$\binom{n}{r} = \frac{n!}{(n-r)!\,r!} = \frac{n!}{r!(n-r)!} = \frac{n!}{(n-(n-r))!(n-r)!} = \binom{n}{n-r}.$$

6.2 Pascal's Formula 帕斯卡公式

Pascal's formula, named after the French mathematician Blaise Pascal, relates the value of $\binom{n+1}{r}$ to the values of $\binom{n}{r}$ and $\binom{n}{r-1}$. Specifically, it says that $\binom{n+1}{r} = \binom{n}{r-1} + \binom{n}{r}$; where n and r are positive integers with $r \leq n$. The Pascal's triangle mentioned earlier is the geometric version of Pascal's formula. Each entry in the Pascal's triangle is a value of $\binom{n}{r}$. Pascal's formula translates into the fact that the value in row $n + 1$, column r equals to the sum of the values in row n, column $r - 1$ plus the value in row n, column r. For example, the third row in Fig. 6.1 of the Pascal's triangle corresponds to the expansion $(x + y)^3$ (we start with row 0 and column 0), and the coefficients are 1　3　3　1; it means that

$$\binom{3}{0} = 1, \binom{3}{1} = 3, \binom{3}{2} = 3, \binom{3}{3} = 1.$$

Theorem 6.1 (The Pascal's Formula)　定理6.1（帕斯卡公式）

Let n and r be positive integers with $r \leq n$, then $\binom{n+1}{r} = \binom{n}{r} + \binom{n}{r-1}$.

当 n 和 r 是正整数，且 $r \leq n$ 时，则 $\binom{n+1}{r} = \binom{n}{r} + \binom{n}{r-1}$.

Proof 证明

$$\binom{n}{r} + \binom{n}{r-1} = \frac{n!}{(n-r)!\,r!} + \frac{n!}{(n-r+1)!\,(r-1)!} = \frac{n!}{(n-r)!\,r!} \cdot \frac{n-r+1}{n-r+1} + \frac{n!}{(n-r+1)!\,(r-1)!} \cdot \frac{r}{r}$$

$$= \frac{n\cdot n! - r\cdot n! + n! + r\cdot n!}{(n-r+1)!\,r!} = \frac{n\cdot n! + n!}{(n-r+1)!\,r!} = \frac{(n+1)\cdot n!}{(n-r+1)!\,r!} = \frac{(n+1)!}{(n-r+1)!\,r!}$$

$$= \frac{(n+1)!}{((n+1)-r)!\,r!} = \binom{n+1}{r}.$$

Example 6.6

Use Pascal's triangle to compute the values of $\binom{5}{2}$ and $\binom{6}{2}$.

Solution

$\binom{5}{2}$ means the value in row 5, column 2 which is equal to the sum of $\binom{4}{2} + \binom{4}{1} = 6 + 4 = 10$.

$\binom{6}{2}$ means the value in row 6, column 2 which is equal to the sum of $\binom{5}{2} + \binom{5}{1} = 10 + 5 = 15$. ∎

Example 6.7

Use Pascal's formula to prove $\binom{n+2}{r+1} = \binom{n}{r+1} + 2\binom{n}{r} + \binom{n}{r-1}$.

Solution

$\binom{n}{r+1} + 2\binom{n}{r} + \binom{n}{r-1} = \binom{n}{r+1} + \binom{n}{r} + \binom{n}{r} + \binom{n}{r-1} = \binom{n+1}{r+1} + \binom{n+1}{r} = \binom{n+2}{r+1}$. ∎

Example 6.8

Use Pascal's formula to prove $\binom{n}{n-2} = \binom{n-1}{n-2} + \binom{n-2}{n-3} + \binom{n-3}{n-4} + \cdots + \binom{3}{2} + \binom{2}{1} + \binom{1}{0}$, then find the value of $\binom{1}{0} + \binom{2}{1} + \binom{3}{2} + \cdots + \binom{13}{12}$.

Solution

$\binom{1}{0} + \binom{2}{1} + \binom{3}{2} + \cdots + \binom{n-3}{n-4} + \binom{n-2}{n-3} + \binom{n-1}{n-2}$

$= \left[\binom{2}{0} + \binom{2}{1}\right] + \binom{3}{2} + \cdots + \binom{n-3}{n-4} + \binom{n-2}{n-3} + \binom{n-1}{n-2} = \left[\binom{3}{1} + \binom{3}{2}\right] + \cdots + \binom{n-3}{n-4} + \binom{n-2}{n-3} + \binom{n-1}{n-2}$

$= \binom{4}{2} + \binom{4}{3} + \cdots + \binom{n-3}{n-4} + \binom{n-2}{n-3} + \binom{n-1}{n-2} = \cdots = \left[\binom{n-2}{n-4} + \binom{n-2}{n-3}\right] + \binom{n-1}{n-2} = \binom{n-1}{n-3} + \binom{n-1}{n-2}$

$= \binom{n}{n-2}\binom{1}{0} + \binom{2}{1} + \binom{3}{2} + \cdots + \binom{13}{12} = \binom{14}{12} = \frac{14!}{2!12!} = \frac{14 \times 13}{2} = 91$. ∎

By now, we have all the required prerequisites; so we can introduce the binomial theorem.

Chapter 6　Binomial Theorem　第六章　二项式定理

Theorem 6.2 (The Binomial Theorem)　定理 6.2（二项式定理）

The binomial theorem states that
$$(a+b)^n = \binom{n}{0}a^n b^0 + \binom{n}{1}a^{n-1}b^1 + \binom{n}{2}a^{n-2}b^2 + \cdots + \binom{n}{n-1}a^1 b^{n-1} + \binom{n}{n}a^0 b^n,$$
where $\binom{n}{r}$ is the binomial coefficient of $a^{n-r}b^r$, and n and r are non-negative integers with $r \leq n$.

根据二项式定理，$(a+b)^n = \binom{n}{0}a^n b^0 + \binom{n}{1}a^{n-1}b^1 + \binom{n}{2}a^{n-2}b^2 + \cdots + \binom{n}{n-1}a^1 b^{n-1} + \binom{n}{n}a^0 b^n$，其中，$\binom{n}{r}$ 是 $a^{n-r}b^r$ 的二项式系数，n 和 r 是非负整数且 $r \leq n$。

Proof　证明

We can use Mathematical Induction or Taylor's Series to prove the theorem. We skip the proof for now, leave it for A-level Further Math Course.

Example 例 6.9

Expand the following expressions:

(a) $(x+2y)^4$;　　(b) $(3a-2b)^5$;　　(c) $\left(2x^2 - \dfrac{1}{x}\right)^3$.

Solution 解

(a) $(x+2y)^4 = x^4 + \binom{4}{1}x^3(2y) + \binom{4}{2}x^2(2y)^2 + \binom{4}{3}x^1(2y)^3 + (2y)^4 = x^4 + 8x^3 y + 24x^2 y^2 + 32xy^3 + 16y^4$.

(b) $(3a-2b)^5 = (3a)^5 + \binom{5}{1}(3a)^4(-2b) + \binom{5}{2}(3a)^3(-2b)^2 + \binom{5}{3}(3a)^2(-2b)^3$
$\qquad\qquad + \binom{5}{4}(3a)^1(-2b)^4 + (-2b)^5$
$\qquad = 243a^5 - 810a^4 b + 1\,080a^3 b^2 - 720a^2 b^3 + 240ab^4 - 32b^5$.

(c) $\left(2x^2 - \dfrac{1}{x}\right)^3 = (2x^2)^3 + \binom{3}{1}(2x^2)^2\left(-\dfrac{1}{x}\right) + \binom{3}{2}(2x^2)^1\left(-\dfrac{1}{x}\right)^2 + \left(-\dfrac{1}{x}\right)^3 = 8x^6 - 12x^3 - \dfrac{1}{x^3} + 6$.

Example 例 6.10

Find the simplified expression of $\binom{n}{0} + \binom{n}{1} + \binom{n}{2} + \cdots + \binom{n}{n}$.

Solution 解

$\binom{n}{0} + \binom{n}{1} + \binom{n}{2} + \cdots + \binom{n}{n} = \binom{n}{0}1^n \cdot 1^0 + \binom{n}{1}1^{n-1} \cdot 1^1 + \binom{n}{2}1^{n-2} \cdot 1^2 + \cdots + \binom{n}{n}1^0 \cdot 1^n = (1+1)^n = 2^n$.

Example 6.11

Find the first four terms in the expansion of $(-3x+2)^6$, in ascending power of x.

Solution

We want ascending power of x, so we rewrite the expansion as $(2-3x)^6$.

$$(2-3x)^6 = 2^6 + \binom{6}{1}(2)^5(-3x)^1 + \binom{6}{2}(2)^4(-3x)^2 + \binom{6}{3}(2)^3(-3x)^3 + \cdots.$$

So the first 4 terms are $64 - 576x + 2160x^2 - 4320x^3$ in ascending power of x.

Example 6.12

Find the first three terms in the expansion of $\left(1 + \dfrac{3x}{2}\right)^6$, in descending power of x.

Solution

We want descending power of x, so we rewrite the expansion as $\left(\dfrac{3x}{2} + 1\right)^6$.

$$\left(\dfrac{3x}{2} + 1\right)^6 = \left(\dfrac{3x}{2}\right)^6 + \binom{6}{1}\left(\dfrac{3x}{2}\right)^5 \cdot 1^1 + \binom{6}{2}\left(\dfrac{3x}{2}\right)^4 \cdot 1^2 + \cdots + \binom{6}{6}1^6.$$

So the first 3 terms are $\dfrac{729}{64}x^6 + \dfrac{729}{16}x^5 + \dfrac{1215}{16}x^4$ in descending power of x.

Example 6.13

Determine whether or not there is a term that is independent of x in the expansion of $\left(x^3 - \dfrac{1}{x}\right)^{10}$.

Solution

Suppose the term that is independent of x is $\binom{10}{n}(x^3)^{10-n}\left(\dfrac{1}{x}\right)^n = k \cdot x^0$, where k is a constant.

$x^{30-3n} \cdot x^{-n} = x^0 \Rightarrow 30 - 4n = 0 \Rightarrow n = \dfrac{30}{4}$ which is not an integer.

Hence, there is no constant term in the expansion.

Example 6.14

Find the term that is independent of x in the expansion of $\left(x^2 - \dfrac{2}{x}\right)^{15}$.

Solution

Suppose the term that is independent of x is $\binom{15}{n}(x^2)^{15-n}\left(\dfrac{-2}{x}\right)^n = k \cdot x^0$, where k is a constant.

$x^{30-2n} \cdot x^{-n} = x^0 \Rightarrow 30 - 3n = 0 \Rightarrow n = 10.$

So the term that is independent of x is $\binom{15}{10}(x^2)^{15-10}\left(\dfrac{-2}{x}\right)^{10} = \binom{15}{10} \cdot (-2)^{10} = 3\,075\,072$. ∎

Example 例 6.15

(a) Find the coefficient of $x^6 y^4$ in the expansion of $(2x^2 - y)^7$.
(b) Find the coefficient of $x^5 y^6$ in the expansion of $(x + 3y^2)^8$.

Solution 解

(a) From the binomial theorem, we know the term involves $x^6 y^4$ in the expansion of $(2x^2 - y)^7$ is equal to
$\binom{7}{4}(2x^2)^3(y)^4 = \binom{7}{4} \cdot 2^3 x^6 y^4 = 35 \cdot 8 x^6 y^4 = 280 x^6 y^4$. The coefficient is 280.

(b) From the binomial theorem, we know the term involves $x^5 y^6$ in the expansion of $(x + 3y^2)^8$ is equal to
$\binom{8}{3}(x)^5(3y^2)^3 = \binom{8}{3} x^5 \cdot 3^3 \cdot y^6 = 1\,512 x^5 y^6$. The coefficient is $1\,512$. ∎

Example 例 6.16

In the expansion of $(2 + ax)^{20}$, the sum of the coefficients is 1. Find the value of a.

Solution 解

Let $f(x) = (2 + ax)^{20}$, the sum of coefficients just equal to $f(1)$.
$(2 + a)^{20} = 1 \Rightarrow 2 + a = \pm 1 \Rightarrow a = -3 \text{ or } -1$.

Observed that if we have an expansion that is in the form of $(1 + y)^n$ where y is any algebraic expression.
Then $(1 + y)^n = 1 + \binom{n}{1} y + \binom{n}{2} y^2 + \cdots + y^n$, use this result; it may help us to solve some question easily.

Clearly, if we have $(a + b)^n$; we can rewrite it as $a^n \left(1 + \dfrac{b}{a}\right)^n$. ∎

Example 例 6.17

Find the coefficient of x^5 in the expansion of $(1 + 2x - 3x^2)^6$.

Solution 解

In this expansion, we have three terms, so we can not use binomial theorem directly.
In order to apply binomial theorem, we have to rewrite the expression as $(1 + y)^6$ where $y = (2x - 3x^2)$.
Using the fact that $(1 + y)^n = 1 + \binom{n}{1} y + \binom{n}{2} y^2 + \cdots + y^n$, so we have

$(1 + 2x - 3x^2)^6 = 1 + \binom{6}{1}(2x - 3x^2) + \binom{6}{2}(2x - 3x^2)^2 + \binom{6}{3}(2x - 3x^2)^3 + \cdots + (2x - 3x^2)^6$.

We can further simplify each expression in the form of $(2x - 3x^2)^n = (2x)^n \left(1 - \dfrac{3x}{2}\right)^n$.

Clearly, there is no terms with x^5 when $n \leqslant 2$.

When $n = 3$, $\binom{6}{3}(2x - 3x^2)^3 = \binom{6}{3}(2x)^3\left(1 - \dfrac{3x}{2}\right)^3$, the term with x^5 is $\binom{6}{3}(2x)^3\binom{3}{2}\left(-\dfrac{3x}{2}\right)^2 = 1\,080x^5$.

When $n = 4$, $\binom{6}{4}(2x - 3x^2)^4 = \binom{6}{4}(2x)^4\left(1 - \dfrac{3x}{2}\right)^4$, the term with x^5 is $\binom{6}{4}(2x)^4\binom{4}{1}\left(-\dfrac{3x}{2}\right)^1 = -1\,440x^5$.

When $n = 5$, $\binom{6}{5}(2x - 3x^2)^5 = \binom{6}{5}(2x)^5\left(1 - \dfrac{3x}{2}\right)^5$, the term with x^5 is $\binom{6}{5}(2x)^5\binom{5}{0}\left(-\dfrac{3x}{2}\right)^0 = 192x^5$.

When $n = 6$, it is clear that it has no term with x^5.

Therefore, the coefficient of x^5 is $1\,080 - 1\,440 + 192 = -168$. ∎

Note 注意

There is an easier way to compute the coefficient, since we have three terms in the expansion; so we can use trinomial expansion.

因为展开式有 3 项，所以，可以采用更简单的方法来计算系数，即三项式展开法．

Example 例 ⟨6.18⟩

Given that $(1 - 2x)^5(1 + 3x)^4 = a + bx + cx^2 + \cdots$, find the values of a, b and c.

Solution 解

$(1 - 2x)^5 = 1 + \binom{5}{1}(-2x) + \binom{5}{2}(-2x)^2 + \cdots$.

$(1 + 3x)^4 = 1 + \binom{4}{1}(3x) + \binom{4}{2}(3x)^2 + \cdots$.

So, $a = 1 \times 1 = 1$, $b = 1 \times \binom{4}{1} \cdot 3 + 1 \times \binom{5}{1} \cdot (-2) = -2$,

$c = 1 \times \binom{4}{2} \cdot 3^2 + 1 \times \binom{5}{2} \cdot (-2)^2 + \binom{5}{1} \cdot (-2) \cdot \binom{4}{1} \cdot (3) = -26$. ∎

Example 例 ⟨6.19⟩

Find the coefficient of x^3 in the expansion of $(1 + 3x - 5x^2)\left(1 + \dfrac{x}{3}\right)^{10}$.

Solution 解

To get x^3, we need the constant term in $(1 + 3x - 5x^2)$ multiply by the term x^3 in the expansion of $\left(1 + \dfrac{x}{3}\right)^{10}$ plus the x term in $(1 + 3x - 5x^2)$ multiply by the term x^2 in the expansion of $\left(1 + \dfrac{x}{3}\right)^{10}$ plus the x^2 term in $(1 + 3x - 5x^2)$ multiply by the term x in the expansion of $\left(1 + \dfrac{x}{3}\right)^{10}$.

$1 \times \binom{10}{3}\left(\dfrac{x}{3}\right)^3 + 3x \cdot \binom{10}{2}\left(\dfrac{x}{3}\right)^2 + (-5x^2) \cdot \binom{10}{1}\left(\dfrac{x}{3}\right)^1 = \dfrac{25}{9}x^3$. ∎

Chapter 6 Binomial Theorem   第六章 二项式定理

Example 例 6.20

If the coefficients of x^4 in the expansion of $(ax+1)^7$ and $(a-x)^9$ are equal. Find the value of a.

Solution 解

The term x^4 in the expansion of $(ax+1)^7$ is $\binom{7}{4}a^4 x^4$.

$(a-x)^9 = a^9\left(1-\dfrac{x}{a}\right)^9$. So, the term x^4 in the expansion of $(a-x)^9$ is $a^9 \binom{9}{4}\left(\dfrac{-x}{a}\right)^4 \cdot \binom{7}{4}a^4 = \binom{9}{4}a^5 \Rightarrow$

$35a^4 = 126a^5 \Rightarrow 35a^4\left(1-\dfrac{18a}{5}\right) = 0$.

Therefore, $a=0$ or $\dfrac{5}{18}$, but $a \neq 0$ otherwise we have no expansion in $(ax+1)^7$.

So $a = \dfrac{5}{18}$ only. ∎

Example 例 6.21

It is given that the ratio of the coefficients of x^3 to the coefficient of x^2 is $4:25$ in the expansion of $\left(\dfrac{5}{2}+\dfrac{1}{5}x\right)^n$. Find the value of n.

Solution 解

$\left(\dfrac{5}{2}+\dfrac{1}{5}x\right)^n = \left(\dfrac{5}{2}\right)^n\left(1+\dfrac{2}{25}x\right)^n$.

So the coefficient of x^3 is $\left(\dfrac{5}{2}\right)^n \binom{n}{3}\left(\dfrac{2}{25}\right)^3$ and the coefficient of x^2 is $\left(\dfrac{5}{2}\right)^n \binom{n}{2}\left(\dfrac{2}{25}\right)^2$.

$\left(\dfrac{5}{2}\right)^n \binom{n}{3}\left(\dfrac{2}{25}\right)^3 : \left(\dfrac{5}{2}\right)^n \binom{n}{2}\left(\dfrac{2}{25}\right)^2 = 4:25 \Rightarrow 4\left(\dfrac{5}{2}\right)^n \binom{n}{2}\left(\dfrac{2}{25}\right)^2 = 25\left(\dfrac{5}{2}\right)^n \binom{n}{3}\left(\dfrac{2}{25}\right)^3 \Rightarrow 4\binom{n}{2} = 25\binom{n}{3}\left(\dfrac{2}{25}\right)$

$\Rightarrow \binom{n}{2} = \dfrac{1}{2}\binom{n}{3} \Rightarrow \dfrac{n!}{(n-2)!2!} = \dfrac{1}{2}\dfrac{n!}{(n-3)!3!}$

$\Rightarrow n(n-1) = \dfrac{n(n-1)(n-2)}{6} \Rightarrow n-2 = 6 \Rightarrow n=8$. ∎

*6.3 Multinomial Expansion 多项式展开

The trinomial expansion is often arisen in STEP or MAT Exam, however it is not in the CAIE A-level Syllabus. This topic is only for students' own interest to read it.

Multinomial Expansion 多项式展开

$(x_1 + x_2 + \cdots + x_k)^n = \sum \dfrac{n!}{n_1! n_2! \cdots n_k!} x_1^{n_1} x_2^{n_2} \cdots x_k^{n_k}$, where $n_1 + n_2 + \cdots + n_k = n$.

When $k = 3$, it is called trinomial expansion. When $k = 2$, it is called binomial expansion.

$(x_1 + x_2 + \cdots + x_k)^n = \sum \dfrac{n!}{n_1! n_2! \cdots n_k!} x_1^{n_1} x_2^{n_2} \cdots x_k^{n_k}$,其中,$n_1 + n_2 + \cdots + n_k = n$.

$k = 3$,为三项式展开;$k = 2$,为二项式展开.

The number of terms in the trinomial expansion $(x + y + z)^n$ is the number of ways to choose n elements with replacement in which order does not matter from a set of 3 elements. The formula is $\binom{3 + n - 1}{n} = \binom{n + 2}{n} = \dfrac{(n + 1)(n + 2)}{2}$.

Example 例 6.22

Find the expansion of $(a + b + c)^n$, for $n = 2, 3$ and 4.

Solution 解

$(a + b + c)^2 = a^2 + b^2 + c^2 + 2ab + 2ac + 2bc$.

$(a + b + c)^3 = a^3 + b^3 + c^3 + 3a^2b + 3a^2c + 3b^2c + 3b^2a + 3c^2a + 3c^2b + 6abc$.

$(a + b + c)^3 = a^4 + b^4 + c^4 + \dfrac{4!}{3!1!}a^3b + \dfrac{4!}{3!1!}a^3c + \dfrac{4!}{3!1!}b^3a + \dfrac{4!}{3!1!}b^3c + \dfrac{4!}{3!1!}c^3a + \dfrac{4!}{3!1!}c^3b + \dfrac{4!}{1!2!}a^2b^2 + \dfrac{4!}{2!2!}a^2c^2 + \dfrac{4!}{2!2!}b^2c^2 + \dfrac{4!}{2!1!1!}a^2bc + \dfrac{4!}{2!1!1!}ab^2c + \dfrac{4!}{2!1!1!}abc^2$.

Example 例 6.23

Find the coefficient of x^3yz^2 in the expansion of $(x + 2y - 3z)^6$.

Solution 解

$\dfrac{6!}{3!1!2!} x^3 (2y)^1 (-3z)^2 = 1\,080 x^3yz^2$.

So the coefficient is $1\,080$.

Example 例 6.24

Find the coefficient of x^5 in the expansion of $(1 + 2x + 3x^2)^8$.

Solution 解

In the expansion, the terms are in the form of $\dfrac{8!}{p!q!r!}(1)^p(2x)^q(3x^2)^r$, where $p + q + r = 8$.

$$\frac{8!}{p!q!r!}(1)^p(2x)^q(3x^2)^r = \frac{8!}{p!q!r!}2^q 3^r x^{q+2r}.$$

So we want $\begin{cases} q + 2r = 5, \\ p + q + r = 8. \end{cases}$

It is obvious that r can only be 0, 1 and 2.

When $r = 0 \Rightarrow q = 5$, $p = 3$. When $r = 1 \Rightarrow q = 3$, $p = 4$. When $r = 2 \Rightarrow q = 1$, $p = 5$.

Therefore, the coefficient of x^5 is $\frac{8!}{3!5!0!}2^5 3^0 + \frac{8!}{4!3!1!}2^3 3^1 + \frac{8!}{5!1!2!}2^1 3^2 = 11\,536$. ∎

Example 例 6.25 (Example 6.17 Revisited 重做例 6.17)

Find the coefficient of x^5 in the expansion of $(1 + 2x - 3x^2)^6$.

Solution 解

In the expansion, the terms are in the form of $\frac{6!}{p!q!r!}(1)^p(2x)^q(-3x^2)^r$, where $p + q + r = 6$.

$\frac{6!}{p!q!r!}(1)^p(2x)^q(-3x^2)^r = \frac{6!}{p!q!r!}2^q(-3)^r x^{q+2r}$. We want $\begin{cases} q + 2r = 5, \\ p + q + r = 6. \end{cases}$

It is obvious that r can only be 0, 1 and 2.

When $r = 0 \Rightarrow q = 5$, $p = 1$. When $r = 1 \Rightarrow q = 3$, $p = 2$. When $r = 2 \Rightarrow q = 1$, $p = 3$.

Therefore, the coefficient of x^5 is $\frac{6!}{1!5!0!}2^5(-3)^0 + \frac{6!}{2!3!1!}2^3(-3)^1 + \frac{6!}{3!1!2!}2^1(-3)^2 = -168$. ∎

Summary of Key Theories 核心定义总结

(1) For each positive integer n, the quantity n factorial denoted $n!$, is defined to be the product of all consecutive integers from 1 to n. $n! = n(n-1)!$, and $0! = 1$.

(2) The symbol $\binom{n}{r} = {}^nC_r$ which is read as "n choose r", denotes the total numbers of subset of size r that can be chosen from a set of n elements. $\binom{n}{r}$ are often called the binomial coefficients, because they arise in the binomial expansion. $\binom{n}{r} = {}^nC_r = \frac{n!}{(n-r)!r!}$, for $0 \leq r \leq n$.

(3) Let n and r be positive integers with $r \leq n$, then $\binom{n+1}{r} = \binom{n}{r} + \binom{n}{r-1}$.

(4) The binomial theorem states that:

$$(a+b)^n = \binom{n}{0}a^n b^0 + \binom{n}{1}a^{n-1}b^1 + \binom{n}{2}a^{n-2}b^2 + \cdots + \binom{n}{n-1}a^1 b^{n-1} + \binom{n}{n}a^0 b^n,$$ where $\binom{n}{r}$ is the binomial

coefficient of $a^{n-r}b^r$, and n and r are non-negative integers with $r \leq n$.

(5) $(x_1 + x_2 + \cdots + x_k)^n = \sum \dfrac{n!}{n_1! n_2! \cdots n_k!} x_1^{n_1} x_2^{n_2} \cdots x_k^{n_k}$, where $n_1 + n_2 + \cdots + n_k = n$.

When $k = 3$, it is called trinomial expansion. When $k = 2$, it is called binomial expansion.

Sequence and Series
第 7 章 数列和级数

We had already encountered a number sequence at IGCSE level. In this section, we define the term **sequence** informally as a set of elements (numbers) written in a row.

If the sequence is given as a_p, a_{p+1}, a_{p+2}, $\cdots$, a_q; each element a_k (read "a sub k") is called a term in the sequence, k is called an index or subscript. a_p is called the initial term or first term of the sequence, and a_q is called the final term or the last term of the sequence.

An explicit formula or general formula for a sequence is a rule that shows how the values of a_n depends on n. A recurrence formula describes a_n which involves the preceding terms. $\{a_n\}$ represents the sequence that can be generated by using a_n as the nth term.

For examples, $u_n = 3n^2 - 1$ for $n = 1, 2, 3, \cdots$ is an explicit formula for the sequence u_n. So $u_1 = 3(1)^2 - 1 = 2$, $u_2 = 3(2)^2 - 1 = 11$ and so on.

$a_n = 2a_{n-1} + 3$ and $u_0 = 2$ for $n = 1, 2, 3, \cdots$, is a recurrence formula for the sequence a_n. So $a_1 = 2a_0 + 3 = 2(1) + 3 = 5$, $a_2 = 2a_1 + 3 = 2(5) + 3 = 13$ and so on.

We only focus on two types of sequence namely Arithmetic Sequence and Geometric Sequence.

7.1 Arithmetic Sequence 等差数列

A sequence $\{u_n\}$ is an arithmetic sequence if each term differs from the previous term by the same constant, that is $u_n - u_{n-1} = d$; d is called the common difference.

The arithmetic sequence is also called an arithmetic progression.

Definition 定义

A sequence $\{u_n\}$ is called an arithmetic sequence (progression), if it satisfies $u_n - u_{n-1} = d$ for all n listed in the sequence, d is called the common difference.

The general formula for the nth term of an arithmetic sequence $\{u_n\}$ is $u_n = a + (n-1)d$, where a is the first term.

如果对一个数列的所有数 n，$u_n - u_{n-1} = d$ 成立，数列 $\{u_n\}$ 被称为等差数列(级数)，其中，d 被称为相位差.

等差数列 $\{u_n\}$ 的第 n 项的通式是 $u_n = a + (n-1)d$，其中，a 为该数列的首项.

Example 例 7.1

Given an arithmetic progression $-5, -1, 3, \cdots$, find the fifth term and the tenth term.

Solution 解

The common difference $d = -1 - (-5) = 4$, so $u_5 = -5 + (5-1)4 = 11$ and $u_{10} = -5 + (10-1)4 = 31$. ∎

Example 例 7.2

The general formula for an arithmetic progression is $a_n = 3 + 5n$. Find the common difference of the sequence.

Solution 解

$a_1 = 3 + 5(1) = 8$, $a_2 = 3 + 5(2) = 13$; so $d = a_2 - a_1 = 13 - 5 = 5$. ∎

Example 例 7.3

If u_n is the nth term of an arithmetic progression where $u_p = p^2$ and $u_q = q^2$. Find the common difference.

Solution 解

Let the first term of the arithmetic progression be a with common difference d.
$u_p = a + (p-1)d = p^2$ and $u_q = a + (q-1)d = q^2$.
$u_p - u_q = (p-1)d - (q-1)d = p^2 - q^2 \Rightarrow d(p - 1 - q + 1) = p^2 - q^2$
$$\Rightarrow d = \frac{p^2 - q^2}{p - q} = \frac{(p+q)(p-q)}{p-q} = p + q.$$
∎

Example 例 7.4

Given an arithmetic progression $100, 93, 86, \cdots$, starting from which term of the progression it begins to be a negative number?

Solution 解

$d = 93 - 100 = -7$. We want $a_n = 100 + (n-1)(-7) < 0 \Rightarrow 7(n-1) > 100 \Rightarrow n > 1 + \dfrac{100}{7} = 15\dfrac{2}{7}$.

So, starting from 16th term, the progression is negative.

Definition 定义

If a, m and b are any three consecutive terms of an arithmetic sequence u_n, then m is called the arithmetic mean of a and b, where $m = \dfrac{a+b}{2}$, it's because $m - a = b - m \Rightarrow 2m = a + b$.

如果 a, m 和 b 是等差数列 u_n 的任意3个连续项，则称 m 为 a 和 b 的算术平均数，其中，$m = \dfrac{a+b}{2}$，因为 $m - a = b - m \Rightarrow 2m = a + b$.

Example 例 7.5

Three numbers formed an arithmetic sequence, their sum is 21 and their product is 231. Find the numbers.

Solution 解

Let the numbers be $a - d$, a, $a + d$.
$a - d + a + a + d = 21 \Rightarrow 3a = 21 \Rightarrow a = 7$.
$(a-d)a(a+d) = 231 \Rightarrow a(a^2 - d^2) = 231$, since $a = 7 \Rightarrow 7(7^2 - d^2) = 231 \Rightarrow d^2 = 16 \Rightarrow d = \pm 4$.
So the three numbers are 3, 7 and 11.

Example 例 7.6

If a, b and c forms an arithmetic progression, show the equation $(b-c)x^2 + (c-a)x + (a-b) = 0$ has equal roots.

Solution 解

Since a, b and c forms an arithmetic progression $\Rightarrow \dfrac{a+c}{2} = b \Rightarrow a + c = 2b$.

The discriminant $D = (c-a)^2 - 4(a-b)(b-c) = c^2 - 2ac + a^2 - 4ab + 4ac - 4bc + 4b^2$
$= a^2 + 4b^2 + c^2 - 4ab + 2ac - 4bc = (a - 2b + c)^2 = 0$.
Recall that $(x + y + z)^2 = x^2 + y^2 + z^2 + 2xy + 2yz + 2xz$.
Hence, the equation has equal root.

Example 例 7.7

An arithmetic progression has first term $2x$ and the second term is x^2. If the common difference for this progression is 3. Find the possible values of x.

Solution 解

$x^2 - 2x = 3 \Rightarrow x^2 - 2x - 3 = 0 \Rightarrow (x - 3)(x + 1) = 0 \Rightarrow x = 3, x = -1$. Now check with the condition $d = 3$.
When $x = 3$, $3^2 - 2 \times 3 = 3$. When $x = -1$, $(-1)^2 - 2(-1) = 3$.
So $x = -1$ or $x = 3$.

Example 例 7.8

Given that x, y and z forms an arithmetic progression. Prove that：
(a) $y+z$, $x+z$, $x+y$ forms an arithmetic progression；
(b) $x^2(y+z)$, $y^2(x+z)$, $z^2(x+y)$ forms an arithmetic progression.

Solution 解

(a) Since x, y and z forms an arithmetic progression with common difference d, this implies that $-x$, $-y$ and $-z$ forms an arithmetic progression with common difference $-d$.
Therefore, $-x+(x+y+z)$, $-y+(x+y+z)$ and $-z+(x+y+z)$ also forms an arithmetic progression with common difference $-d$. Hence, $y+z$, $x+z$, $x+y$ forms an arithmetic progression.

(b) x, y and z forms an arithmetic progression. $x+z = 2y$.
$$x^2(y+z) + z^2(x+y) = x^2y + x^2z + xz^2 + yz^2 = y(x^2+z^2) + xz(x+z) = y(x^2+z^2) + 2xyz$$
$$= y(x^2+z^2+2xz) = y(x+z)^2 = y \cdot 2y \cdot (x+z) = 2y^2(x+z).$$
Hence, $x^2(y+z)$, $y^2(x+z)$, $z^2(x+y)$ forms an arithmetic progression. ∎

Note 注意

If x, y and z forms an arithmetic progression, then kx, ky and kz forms an arithmetic progression with nonzero constant k. Also $x+k$, $y+k$ and $z+k$ forms an arithmetic progression as well.
如果 x, y 和 z 构成等差数列，则 kx, ky 和 kz 构成非零常数 k 的等差数列. 同时, $x+k$, $y+k$, 和 $z+k$ 也构成一个等差数列.

7.2 Arithmetic Series 等差级数

A series is the sum of the terms of a sequence, we usually denoted it by S_n.
For a sequence $\{a_n\}$, $S_n = a_1 + a_2 + \cdots + a_n$.
For an arithmetic sequence with first term a and common difference d：
$$S_n = a_1 + a_2 + \cdots + a_n = a + (a+d) + (a+2d) + \cdots + [a+(n-1)d].$$
If we reverse the order, we get
$$S_n = a_n + a_{n-1} + \cdots + a_2 + a_1 = [a+(n-1)d] + [a+(n-2)d] + \cdots + (a+d) + a.$$
We sum these two equations up, we would obtain：
$$2S_n = [2a+(n-1)d] + [2a+(n-1)d] + \cdots + [2a+(n-1)d] = n \cdot [2a+(n-1)d] = n \cdot (a_1+a_n)$$
$$\Rightarrow S_n = \frac{(a_1+a_n)n}{2} = \frac{[2a+(n-1)d]n}{2}.$$
When we are given an expression of S_n,
$$S_1 = a_1, \ S_2 = a_1+a_2, \ S_{n-1} = a_1+a_2+\cdots+a_{n-1}, \ S_n = a_1+a_2+\cdots+a_{n-1}+a_n \Rightarrow S_n - S_{n-1} = a_n,$$
the nth term of the sequence.

Chapter 7　Sequence and Series　　第七章　数列和级数

> **Theorem 7.1　定理 7.1**
>
> For an arithmetic sequence $\{a_n\}$, the series $S_n = a_1 + a_2 + \cdots + a_{n-1} + a_n = \dfrac{(a_1 + a_n)n}{2}$.
>
> 对于等差数列 $\{a_n\}$，级数 $S_n = a_1 + a_2 + \cdots + a_{n-1} + a_n = \dfrac{(a_1 + a_n)n}{2}$.

Proof　证明

It is given above. ∎

Example 例 7.9

Find an expression of S_n for an arithmetic sequence with first term a and common difference d and expressed it as a quadratic equation of n.

Solution 解

Let the sequence be $\{u_n\}$, so $u_1 = a$ and $u_n = a + (n-1)d$.

Hence, $S_n = \dfrac{(a + a + (n-1)d)n}{2} = \dfrac{(2a + (n-1)d)}{2} n = \dfrac{d}{2} n^2 + \dfrac{(2a-d)}{2} n$. ∎

Example 例 7.10

The arithmetic series is given by $S_n = 6n^2 - 4n$.

(a) Find the first term and the common difference of the arithmetic progression.

(b) Find the explicit formula for the nth term of the sequence u_n.

Solution 解

(a) $S_1 = 6(1)^2 - 4(1) = 2 = u_1$, $S_2 = 6(2)^2 - 4(2) = 16 = u_1 + u_2 \Rightarrow u_2 = S_2 - S_1 = 16 - 2 = 14$.

$d = u_2 - u_1 = 14 - 2 = 12$.

So the first term is 2, and the common difference is 12.

(b) **Method 1**

$u_n = S_n - S_{n-1} = 6n^2 - 4n - [6(n-1)^2 - 4(n-1)] = 6n^2 - 4n - (6n^2 - 16n + 10) = 12n - 10$.

Method 2

From (a) we know the first term is 2 common difference is 12, so $u_n = 2 + (n-1)12 = 12n - 10$. ∎

Example 例 7.11

The sum of the 7th term to 11th term of an arithmetic progression is 100, and the sum of 12th term to 18th term of the same progression is 224. Find the sum of 2nd term to 5th term.

Solution 解

Let $\{u_n\}$ be the arithmetic sequence with first term a and common difference d.

From the question, we know $S_{11} - S_6 = 100$ and $S_{18} - S_{11} = 224$.

Since these are arithmetic series, so $S_{11} - S_6 = \dfrac{(u_{11} + u_7)5}{2} = \dfrac{5}{2}(a + 10d + a + 6d) = 100$; and

$S_{18} - S_{11} = \dfrac{(u_{18} + u_{12})7}{2} = \dfrac{7}{2}(a + 17d + a + 11d) = 224$.

$\begin{cases} a + 8d = 20 \\ a + 14d = 32 \end{cases} \Rightarrow a = 4, \ d = 2.$

Therefore, $S_5 - S_1 = \dfrac{(u_5 + u_2)4}{2} = \dfrac{4}{2}(a + 4d + a + d) = 36$. ∎

7.3 Geometric Sequence 等比数列

Definition 定义

A sequence $\{u_n\}$ is called a geometric sequence (progression), if it satisfied $\dfrac{u_n}{u_{n-1}} = r$ for all n listed in the sequence, r is called the common ratio.

The general formula for n^{th} term of a geometric sequence $\{u_n\}$ is $u_n = ar^{n-1}$, where a is the first term.

如果该数列满足 $\dfrac{u_n}{u_{n-1}} = r$, 称数列 $\{u_n\}$ 为等比数列(级数), 其中, r 被称为公比.

等比数列 $\{u_n\}$ 第 n 项通式为 $u_n = ar^{n-1}$, 其中, a 为首项.

Example 例 7.12

Given a geometric sequence $18, 9, \dfrac{9}{2}, \cdots$, find the expression of the nth term, and hence, find the 5th term and 7th term.

Solution 解

Let $\{u_n\}$ be the geometric sequence. The common ratio $r = \dfrac{9}{18} = \dfrac{1}{2}$.

So the nth term $u_n = 18\left(\dfrac{1}{2}\right)^{n-1}$.

$u_5 = 18\left(\dfrac{1}{2}\right)^{5-1} = 18\left(\dfrac{1}{2}\right)^4 = \dfrac{9}{8}$ and $u_7 = 18\left(\dfrac{1}{2}\right)^{7-1} = 18\left(\dfrac{1}{2}\right)^6 = \dfrac{9}{32}$. ∎

> **Definition 定义**
>
> If a, m and b are any three consecutive terms of a geometric sequence u_n, then m is called the geometric mean of a and b, where $m = \pm \sqrt{ab}$, it's because $\dfrac{m}{a} = \dfrac{b}{m} \Rightarrow m^2 = ab$.
>
> 如果 a, m 和 b 为等比数列 u_n 的任意3个连续项，那么，称 m 为 a 和 b 的等比平均数，其中，$m = \pm \sqrt{ab}$，因为 $\dfrac{m}{a} = \dfrac{b}{m} \Rightarrow m^2 = ab$.

Example 例 7.13

The sequence $3, -6, x, y$ is such that the first three terms form an arithmetic progression and the last three terms form a geometric progression. Find the values of x and y.

Solution 解

$3, -6, x$ forms an arithmetic progression, so $\dfrac{3+x}{2} = -6 \Rightarrow x - 15$.

$-6, -15, y$ forms a geometric progression, so $-6y = 15^2 \Rightarrow y = \dfrac{-75}{2}$. ∎

Example 例 7.14

Find the expression of the nth term of the arithmetic progression $\{u_n\}$ with first term 2 and a nonzero common difference d, whose second, eleventh, and thirty-first terms are the first three terms of a geometric progression.

Solution 解

Let the first term of the geometric progression be a, and common ratio r.
Therefore, $2 + d = a$, $2 + 10d = ar$ and $2 + 30d = ar^2$.

$2 + 10d = ar \Rightarrow 2 + 10d = (2 + d)r$. \hfill (1)

$2 + 30d = ar^2 \Rightarrow 2 + 30d = (2 + 10d)r$. \hfill (2)

$(2) - (1) \Rightarrow 20d = 9dr$, since $d \neq 0 \Rightarrow r = \dfrac{20}{9}$.

Therefore, $2 + 10d = \dfrac{20}{9}a = \dfrac{20}{9}(2 + d) \Rightarrow d = \dfrac{22}{70}$; so $u_n = 2 + (n-1)\dfrac{22}{70}$. ∎

Example 例 7.15

Given x, y, z, w forms a geometric progression.
Prove that $x + y, y + z, z + w$ forms a geometric progression.

Solution 解

x, y, z, w forms a geometric progression $\Rightarrow xz = y^2$, $yw = z^2$ and $xw = yz$.
Now, $(x + y)(z + w) = xz + xw + yz + yw = y^2 + yz + yz + z^2 = y^2 + 2yz + z^2 = (y + z)^2$.
Therefore, $x + y, y + z, z + w$ forms a geometric progression. ∎

7.4 Geometric Series 等比级数

For a geometric sequence with first term a and common ratio r, then
$S_n = a_1 + a_2 + a_3 + \cdots + a_n = a + ar + ar^2 + \cdots + ar^{n-1}$; if we multiply S_n by r, we get
$rS_n = ar + ar^2 + \cdots + ar^{n-1} + ar^n \Rightarrow rS_n - S_n = ar^n - a \Rightarrow S_n = \dfrac{a(r^n - 1)}{r - 1} = \dfrac{a(1 - r^n)}{1 - r}$.

> **Theorem 7.2 定理 7.2**
>
> For a geometric sequence $\{a_n\}$ with first term a and common ratio r, then the series
> $$S_n = a_1 + a_2 + \cdots + a_{n-1} + a_n = S_n = \dfrac{a(r^n - 1)}{r - 1} = \dfrac{a(1 - r^n)}{1 - r}.$$
> 如果等比数列 $\{a_n\}$ 的第一项为 a，公比为 r，则该等比数列的级数为
> $$S_n = a_1 + a_2 + \cdots + a_{n-1} + a_n = S_n = \dfrac{a(r^n - 1)}{r - 1} = \dfrac{a(1 - r^n)}{1 - r}.$$

Proof 证明

It is given above.

Example 例 7.16

Find the sum of the geometric sequence $2, 6, 18, \cdots$ up to and includes the 10th term.

Solution 解

$r = \dfrac{6}{2} = 3$, $S_{10} = \dfrac{2(3^{10} - 1)}{3 - 1} = 59\,048$.

Example 例 7.17

Find a formula for S_n, the sum of the first n terms of the series $8 - 4 + 2 - 1 + \cdots$.

Solution 解

The series is a geometric series because $\dfrac{-4}{8} = \dfrac{2}{-4} = \dfrac{-1}{2}$, so it is a geometric series with common ratio $r = \dfrac{-1}{2}$. $S_n = \dfrac{a(1 - r^n)}{1 - r} = \dfrac{8\left[1 - \left(\dfrac{-1}{2}\right)^n\right]}{1 - \left(\dfrac{-1}{2}\right)} = \dfrac{8\left[1 - \left(\dfrac{-1}{2}\right)^n\right]}{\dfrac{3}{2}} = \dfrac{16}{3}\left[1 - \left(\dfrac{-1}{2}\right)^n\right]$.

Chapter 7 Sequence and Series

Example 7.18

Find the coefficient of x^7 in the expansion of $(1 + 2x) + (1 + 2x)^2 + \cdots + (1 + 2x)^{10}$.

Solution

$(1 + 2x) + (1 + 2x)^2 + \cdots + (1 + 2x)^{10} = \dfrac{(1 + 2x)[(1 + 2x)^{10} - 1]}{(1 + 2x) - 1} = \dfrac{(1 + 2x)^{11} - (1 + 2x)}{2x}$.

The coefficient in x^7 in the expansion is just equal to the coefficient of x^8 in $\dfrac{(1 + 2x)^{11}}{2}$.

So it is $\dfrac{1}{2}\dbinom{11}{8}(2)^8 = 21\,120$.

Example 7.19

Find the expression of $9 + 99 + 999 + 9\,999 + \cdots + n9\text{'s}$, and then deduce the formula for $k + kk + kkk + k\,kkk + \cdots + nk\text{'s}$ for k being the positive integer from 1 to 9 inclusive.

Solution

$9 + 99 + 999 + 9\,999 + \cdots + n9\text{'s} = (10 - 1) + (100 - 1) + (1\,000 - 1) + \cdots + (10^n - 1)$

$\qquad = 10 + 100 + 1\,000 + \cdots + 10^n - (1 + 1 + 1 + \cdots + 1)$

$\qquad = \dfrac{10(10^n - 1)}{10 - 1} - n = \dfrac{10(10^n - 1)}{9} - n$.

$\dfrac{9}{k}(k + kk + kkk + k\,kkk + \cdots + nk\text{'s}) = 9 + 99 + 999 + 9\,999 + \cdots + n9\text{'s} = \dfrac{10(10^n - 1)}{9} - n$

$\Rightarrow k + kk + kkk + k\,kkk + \cdots + nk\text{'s} = \dfrac{k}{9}\left[\dfrac{10(10^n - 1)}{9} - n\right]$.

Example 7.20

Find the expression of $0.9 + 0.99 + 0.999 + \cdots + 0.99\cdots 99$, where there is n's 9 in the last term; and then deduce the formula for $0.k + 0.kk + 0.kkk + \cdots + 0.kk\cdots kk$, where there is n's k in the last term for k being the positive integer from 1 to 9 inclusive.

Solution

$0.9 + 0.99 + 0.999 + \cdots + 0.99\cdots 99 = 1 - \dfrac{1}{10} + 1 - \dfrac{1}{10^2} + 1 - \dfrac{1}{10^3} + \cdots + 1 - \dfrac{1}{10^n}$

$\qquad = n - \left(\dfrac{1}{10} + \dfrac{1}{10^2} + \cdots + \dfrac{1}{10^n}\right)$

$\qquad = n - \dfrac{\dfrac{1}{10}\left(1 - \dfrac{1}{10^n}\right)}{1 - \dfrac{1}{10}} = n - \dfrac{1}{9}\left(1 - \dfrac{1}{10^n}\right)$.

$0.k + 0.kk + 0.kkk + \cdots + 0.kk\cdots kk = \dfrac{k}{9}(0.9 + 0.99 + 0.999 + \cdots + 0.99\cdots 99) = \dfrac{k}{9}\left[n - \dfrac{1}{9}\left(1 - \dfrac{1}{10^n}\right)\right]$.

7.5　Infinite Geometric Series　无穷等比级数

If $\{u_n\}$ is an infinite geometric sequence, we define $S_\infty = u_1 + u_2 + \cdots$ up to infinity. If S_∞ equals to a finite number, then we called the series is convergent; otherwise it is divergent.

We are not going to prove the result for infinite geometric series in this section, however, we will prove it later when we have the ideas of "limit".

If the common ratio r of the infinite geometric sequence satisfied $|r| \geq 1$, then the infinite series is divergent. If $|r| < 1$, then the infinite series is convergent and $S_\infty = \dfrac{u_1}{1-r}$.

Theorem 7.3　定理 7.3

For an infinite geometric sequence $\{a_n\}$ with first term a and common ratio r, then the infinite series converges when $|r| < 1$, and $S_\infty = \dfrac{a}{1-r}$. It is divergent when $|r| \geq 1$.

如果无穷等比数列 $\{a_n\}$，第一项为 a，且公比为 r，那么，当 $|r| < 1$ 时，该无穷级数收敛，并且 $S_\infty = \dfrac{a}{1-r}$；当 $|r| \geq 1$ 时，该级数发散。

Proof　证明

We skip the proof for now.

Example 例 7.21

Find the sum of the following geometric series.

(a) $1 - \dfrac{1}{2} + \dfrac{1}{4} - \dfrac{1}{8} + \cdots$;

(b) $3 + \dfrac{3}{4} + \dfrac{3}{4^2} + \dfrac{3}{4^3} + \cdots$;

(c) $2 + 4 + 8 + 16 + \cdots$;

(d) $-3 - 6 - 12 - 24 - \cdots$;

(e) $2 - r + 8 - 16 + \cdots$;

(f) $-2 + 4 - 8 + 16 + \cdots$.

Solution　解

(a) $r = \dfrac{-1}{2}$, so $S_\infty = \dfrac{1}{1 - \left(\dfrac{-1}{2}\right)} = \dfrac{1}{\dfrac{3}{2}} = \dfrac{2}{3}$.

(b) $r = \dfrac{1}{4}$, so $S_\infty = \dfrac{3}{1 - \left(\dfrac{1}{4}\right)} = \dfrac{3}{\dfrac{3}{4}} = 4$.

(c) $r = 2$ and $a = 2$, so $S_\infty = \infty$; it is divergent.
(d) $r = 2$ and $a = -3$, so $S_\infty = -\infty$; it is divergent.
(e) $r = -2$ and $a = 2$, so S_∞ does not exist; it is divergent.
(f) $r = -2$ and $a = -2$, so S_∞ does not exist; it is divergent.

Example 7.22

Simplify the infinite series $1 + x + x^2 + x^3 + \cdots$ for $|x| < 1$.

Solution

$$1 + x + x^2 + x^3 + \cdots = \frac{1}{1-x} = (1-x)^{-1}.$$

Example 7.23

Simplify the infinite series $1 + 2x + 3x^2 + \cdots + nx^{n-1} + \cdots$ for $|x| < 1$.

Solution

Suppose $S = 1 + 2x + 3x^2 + \cdots + nx^{n-1} + \cdots \Rightarrow xS = x + 2x^2 + 3x^3 + \cdots$

$$\Rightarrow (1-x)S = 1 + x + x^2 + x^3 + \cdots = \frac{1}{1-x}.$$

So $S = \dfrac{1}{(1-x)^2} = (1-x)^{-2}$.

Example 7.24

Simplify the infinite series $1 - x + x^2 - x^3 + \cdots + (-x)^{n-1} \cdots$ for $|x| < 1$.

Solution

$$1 - x + x^2 - x^3 + \cdots + (-x)^{n-1} \cdots = \frac{1}{1-(-x)} = \frac{1}{1+x}.$$

Example 7.25

Simplify the infinite series $1 - 2x + 3x^2 - 4x^3 + \cdots + n(-x)^{n-1} + \cdots$ for $|x| < 1$.

Solution

Suppose $S = 1 - 2x + 3x^2 - 4x^3 + \cdots + n(-x)^{n-1} + \cdots$

$$\Rightarrow xS = x - 2x^2 + 3x^3 + \cdots + n(-1)^{n-1}(x)^n + \cdots$$

$$\Rightarrow (1+x)S = 1 - x + x^2 - x^3 + \cdots = \frac{1}{1+x}.$$

So $S = \dfrac{1}{(1+x)^2} = (1+x)^{-2}$.

Example 7.26

Find the rational number which represents the repeated decimal $0.\overline{32}$.

Solution

Method 1

Using Infinite Geometric Series.

$$0.\overline{32} = 0.32 + 0.0032 + 0.000032 + \cdots = \frac{0.32}{1 - \frac{1}{100}} = \frac{\frac{32}{100}}{\frac{99}{100}} = \frac{32}{99}.$$

Method 2

Using Algebra.

Let $x = 0.\overline{32} \Rightarrow 100x = 32.\overline{32}$.

$100x - x = 32.\overline{32} - 0.\overline{32} = 32 \Rightarrow 99x = 32 \Rightarrow x = \frac{32}{99}$.

Example 7.27

Find the value of $\dfrac{2}{15} + \dfrac{34}{15^2} + \dfrac{98}{15^3} + \cdots + \dfrac{5^n + (-3)^n}{15^n} + \cdots$.

Solution

$$\frac{2}{15} + \frac{34}{15^2} + \frac{98}{15^3} + \cdots + \frac{5^n + (-3)^n}{15^n} + \cdots$$

$$= \left(\frac{1}{3} - \frac{1}{5}\right) + \left(\frac{5^2 + (-3)^2}{15^2}\right) + \left(\frac{5^3 + (-3)^3}{15^3}\right) + \cdots + \frac{5^n + (-3)^n}{15^n} + \cdots$$

$$= \left(\frac{1}{3} - \frac{1}{5}\right) + \left(\frac{1}{3^2} + \frac{1}{5^2}\right) + \left(\frac{1}{3^3} - \frac{1}{5^3}\right) + \cdots + \left(\frac{1}{3^n} + \frac{(-1)^n}{5^n}\right) + \cdots$$

$$= \left(\frac{1}{3} + \frac{1}{3^2} + \frac{1}{3^3} + \cdots + \frac{1}{3^n} + \cdots\right) + \left(\frac{-1}{5} + \frac{1}{5^2} - \frac{1}{5^3} + \cdots + \frac{(-1)^n}{5^n} + \cdots\right)$$

$$= \frac{\frac{1}{3}}{1 - \frac{1}{3}} + \frac{\left(-\frac{1}{5}\right)}{1 - \left(-\frac{1}{5}\right)} = \frac{1}{2} + \frac{-1}{6} = \frac{1}{3}.$$

Example 7.28

The second term and fifth term of a geometric sequence are $\dfrac{1}{3}$ and $\dfrac{8}{81}$ respectively. Find the sum to infinity of the sequence.

Solution

Let the geometric sequence $\{u_n\}$ has first term a and common ratio r.

$u_2 = ar = \dfrac{1}{3}$ and $u_5 = ar^4 = \dfrac{8}{81} \Rightarrow \dfrac{u_5}{u_2} = \dfrac{ar^4}{ar} = r^3 = \dfrac{\frac{8}{81}}{\frac{1}{3}} = \dfrac{8}{27} \Rightarrow r = \dfrac{2}{3}.$

$ar = \dfrac{1}{3} \Rightarrow a = \dfrac{1}{2};\ S_\infty = \dfrac{\frac{1}{2}}{1 - \frac{2}{3}} = \dfrac{3}{2}.$ ∎

Summary of Key Theories　核心定义总结

(1) A sequence $\{u_n\}$ is called an arithmetic sequence (progression), if it satisfies $u_n - u_{n-1} = d$ for all n listed in the sequence, d is called the common difference. The general formula for the nth term of an arithmetic sequence $\{u_n\}$ is $u_n = a + (n-1)d$, where a is the first term.

(2) If a, m and b are any three consecutive terms of an arithmetic sequence u_n, then m is called the arithmetic mean of a and b, where $m = \dfrac{a+b}{2}$, it's because $m - a = b - m \Rightarrow 2m = a + b$.

(3) For an arithmetic sequence $\{a_n\}$, the series: $S_n = a_1 + a_2 + \cdots + a_{n-1} + a_n = \dfrac{(a_1 + a_n)n}{2}$.

(4) A sequence $\{u_n\}$ is called a geometric sequence (progression), if it satisfied $\dfrac{u_n}{u_{n-1}} = r$ for all n listed in the sequence, r is called the common ratio. The general formula for n^{th} term of a geometric sequence $\{u_n\}$ is $u_n = ar^{n-1}$, where a is the first term.

(5) If a, m and b are any three consecutive terms of a geometric sequence u_n, then m is called the geometric mean of a and b, where $m = \pm\sqrt{ab}$, it's because $\dfrac{m}{a} = \dfrac{b}{m} \Rightarrow m^2 = ab$.

(6) For a geometric sequence $\{a_n\}$ with first term a and common ratio r, then the series: $S_n = a_1 + a_2 + \cdots + a_{n-1} + a_n = S_n = \dfrac{a(r^n - 1)}{r - 1} = \dfrac{a(1 - r^n)}{1 - r}$.

(7) For an infinite geometric sequence $\{a_n\}$ with first term a and common ratio r, then the infinite series converges when $|r| < 1$, and $S_\infty = \dfrac{a}{1 - r}$. It is divergent when $|r| \geq 1$.

Differentiation

第 8 章 微 分

In this chapter we are going to begin the study of calculus. Calculus is different from the mathematics that you have studied previously. Calculus was created to describe how quantities change. It has two basic procedures that are opposites of one another.
- Differentiation
- Integration

The main objective of differentiation is finding the rate of change of a given quantity, while the main objective of integration is finding the quantity having a given rate of change.

Both of these procedures are based on the concept of the "limit".

Even though the idea of "limit" is not in the CAIE Math Syllabus, but we will begin this chapter with the concept of "limit of a function".

*8.1 Limit 极限

The concept of "limit" is the cornerstone of developing calculus. In examining the limit of a function $f(x)$ at a point $x = a$, we wish to know what the value of $f(x)$ is getting close to as x gets closer to a. To do this, $f(x)$ must be defined at points arbitrarily close to a although not necessarily at a itself.

> **Definition** 关于函数极限的非正式定义
>
> If $f(x)$ is defined for all x near $x = a$, except possibly at $x = a$. If $f(x)$ approaches the value L as x approaches a, then we write $\lim_{x \to a} f(x) = L$.
>
> 如果函数 $f(x)$ 对所有接近 $x = a$ 的 x 成立，当 x 接近 a 时，$f(x)$ 接近值 L，那么，$\lim_{x \to a} f(x) = L$.

$\lim_{x \to a} f(x)$ can sometimes be evaluated and equal to the value of $f(a)$, this is the case that if $f(x)$ is defined in an open interval containing $x = a$ and the graph of $f(x)$ is passing the point $(a, f(a))$ unbrokenly (In mathematics, we said the function $f(x)$ is continuous at $x = a$).

Theorem 8.1   定理 8.1

If $f(x)$ is a polynomial, rational or trigonometric function, then $\lim_{x \to a} f(x) = f(a)$ provided that $x = a$ is in the domain of $f(x)$.

如果 $f(x)$ 是一个多项式有理函数或三角函数，则 $\lim_{x \to a} f(x) = f(a)$ 假定 $x = a$ 在 $f(x)$ 的定义域中．

Proof   证明

The proof is beyond the scope of this course, you can find the proof in most of the Mathematical Analysis textbook.

Example 例 8.1

Evaluate the following limits：

(a) $\lim_{x \to 1} x^2 + 2x + 1$；

(b) $\lim_{x \to 2} \dfrac{3x - 1}{x + 2}$；

(c) $\lim_{x \to \pi} \cos x$.

Solution 解

(a) $\lim_{x \to 1} x^2 + 2x + 1 = 1^2 + 2(1) + 1 = 4$.

(b) $\lim_{x \to 2} \dfrac{3x - 1}{x + 2} = \dfrac{3(2) - 1}{2 + 2} = \dfrac{5}{4}$.

(c) $\lim_{x \to \pi} \cos x = \cos \pi = -1$.

We now consider the function $f(x) = \dfrac{x^2 - 1}{x - 1}$, the function is not defined at $x = 1$, and the graph of the function looks like a straight line $y = x + 1$ with an open hole on the point $(1, 2)$ (Fig. 8.1).

As x approaches to 1, the function $f(x)$ is getting closer to 2, even though the point $(1, 2)$ does not exist. In such case, we said the limit of $f(x)$ as x approaches to 1 is 2.

So, we write $\lim_{x \to 1} \dfrac{x^2 - 1}{x - 1} = 2$.

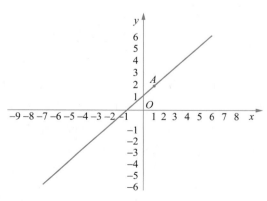

Fig. 8.1

8.2 Limit at Infinity for Rational Function
有理函数在无穷大的极限

The method of finding the end behavior ($x \to \pm\infty$) is to divide the numerator and denominator by the highest degree of x appearing in the denominator. The limit of a rational function at infinity and negative infinity either both exist and equal or both fail to exist.

求 $x \to \pm\infty$ 最终行为的方法是将分子、分母同时除以分母中出现的最高次数 x. 有理函数在正无穷和负无穷处的极限要么都存在且相等,要么都不存在.

Example 例 8.2

Evaluate the followings：

(a) $\lim\limits_{x \to \pm\infty} \dfrac{2x^2 - x + 3}{7x^2 - 4x + 3}$；　(b) $\lim\limits_{x \to \pm\infty} \dfrac{7x - 1}{2x^2 + 5x - 100}$.

Solution 解

(a) $\lim\limits_{x \to \pm\infty} \dfrac{2x^2 - x + 3}{7x^2 - 4x + 3} = \lim\limits_{x \to \pm\infty} \dfrac{2 - \dfrac{1}{x} + \dfrac{3}{x^2}}{7 - \dfrac{4}{x} + \dfrac{3}{x^2}} = \dfrac{2}{7}$.

(b) $\lim\limits_{x \to \pm\infty} \dfrac{7x - 1}{2x^2 + 5x - 100} = \lim\limits_{x \to \pm\infty} \dfrac{\dfrac{7}{x} - \dfrac{1}{x^2}}{2 + \dfrac{5}{x} - \dfrac{100}{x^2}} = 0$.

8.3 Tangent Lines and Gradients　正切线及斜率

Let a curve C be the graph of $f(x)$ and P be a point (x_0, y_0) on C (Fig. 8.2).

Assuming P is not a boundary point of the curve C. What do we mean when we say that the line L is the tangent line to the function $f(x)$ at P? A reasonable definition of tangency can be stated in terms of limits. If Q is any point other than P lies on the curve C, then the line through P and Q is called a secant line.

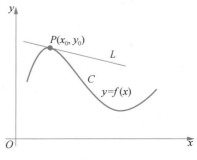

Fig. 8.2

The limit of gradient of the secant line PQ as Q approaches P alone the curve C would be the gradient of the tangent line L.

$P(x_0, f(x_0))$, let $Q(x_0 + h, f(x_0 + h))$; so the gradient of PQ is $\dfrac{f(x_0 + h) - f(x_0)}{h}$ and this expression is known as the "Newton quotient or difference quotient" for $f(x)$ at $x = x_0$.

Suppose the function $f(x)$ is continuous at $x = x_0$ and $\lim\limits_{h \to 0} \dfrac{f(x_0 + h) - f(x_0)}{h} = m$ exists.

Therefore, m is the gradient of the tangent line L passing through $P(x_0, y_0)$ to the graph of $f(x)$.

Then the equation for the tangent line L is $\dfrac{y - y_0}{x - x_0} = m \Rightarrow y = m(x - x_0) + y_0$.

The normal line and the tangent line are perpendicular, therefore the product of the gradient of tangent line and the gradient of the normal line would be -1.

8.4 The Derivative 导数

The gradient is the same at all points for a straight line, however, the gradient may vary at different points for curves other than straight line. Thus, the gradient of a curve at point x is a function of x. If the gradient of tangent line at a point x to $f(x)$ is finite, then we say that $f(x)$ is differentiable at that point. We called the gradient of tangent line of $f(x)$, the derivative of $f(x)$; and the derivative is the limit of the difference quotient.

Definition 定义

The derivative of a function $f(x)$ is another function denoted by $f'(x)$ (read "f prime"), where $f'(x) = \lim\limits_{h \to 0} \dfrac{f(x + h) - f(x)}{h}$ at all points x for which the limit exists.

If $f'(x)$ exist at a point $x = x_0$, we say the function $f(x)$ is differentiable at $x = x_0$.

函数 $f(x)$ 的导数可以用另一个函数 $f'(x)$ 表达，在极限存在的所有点 x 上，$f'(x) = \lim\limits_{h \to 0} \dfrac{f(x + h) - f(x)}{h}$. 如果 $f'(x)$ 对 $x = x_0$ 的点成立，可以说函数 $f(x)$ 在 $x = x_0$ 上是可微的.

The definition of derivative is also known as the first principle formula.

Note 注意

In current CAIE syllabus, the definition of derivative is not covered; therefore you do not need to use this definition to find the derivative in the exam unless the syllabus has changed.

目前 CAIE 大纲并不涉及导数的定义，所以，考生不需要在考试中使用这个定义去求得导数，除非后期大纲有变动.

Example 例 8.3

Use the definition of derivative to find $f'(x)$, where $f(x) = x^2 + 1$; and find the equation of the tangent line and normal line to $f(x)$ at the point $(1, 2)$.

Solution 解

$$f'(x) = \lim_{h \to 0} \frac{f(x+h) - f(x)}{h} = \lim_{h \to 0} \frac{(x+h)^2 + 1 - (x^2 + 1)}{h} = \lim_{h \to 0} \frac{x^2 + 2xh + h^2 + 1 - x^2 - 1}{h}$$

$$= \lim_{h \to 0} \frac{2xh + h^2}{h} = \lim_{h \to 0} \frac{h(2x + h)}{h} = 2x.$$

The gradient of the tangent line at $(1, 2)$ is $f'(1) = 2$.

The equation of the tangent line is $\dfrac{y-2}{x-1} = 2 \Rightarrow y = 2x$.

The gradient of the normal line at $(1, 2)$ is $\dfrac{-1}{f'(1)} = \dfrac{-1}{2}$.

The equation of the normal line is $\dfrac{y-2}{x-1} = \dfrac{-1}{2} \Rightarrow y = \dfrac{-1}{2}x + \dfrac{5}{2}$. ∎

8.5 Other Notations for Derivative 导数的其他符号

If $y = f(x)$, the dependent variable is y while the independent variable is x; so we can denote the derivative of the function with respect to x as the following ways:

$f'(x) = y' = \dfrac{dy}{dx} = \dfrac{d}{dx}(f(x))$, $\dfrac{dy}{dx} = \lim\limits_{\Delta x \to \infty} \dfrac{\Delta y}{\Delta x}$, therefore $f'(x)$ is the instantaneous rate of change of y with respect to x.

The notation $\dfrac{dy}{dx}$ and $\dfrac{d}{dx}(f(x))$ are called the Leibniz notation, name after Gottfried Wilhelm Leibniz, one of the creator of calculus.

The symbol $\dfrac{d}{dx}$ is a differential operator and should be read as "the derivative with respect to x of …".

The value of the derivative of a function at $x = x_0$ can also be expressed in the following ways:

$f'(x_0) = \dfrac{d}{dx} f(x) \Big|_{x = x_0} = y' \Big|_{x = x_0}$.

The symbol " $\Big|_{x = x_0}$ " is called an evaluation symbol, it means the expression preceding should be evaluated at $x = x_0$.

8.6 Basic Differentiation Rules 基本微分法

If we have to calculate every derivative of functions from the definition of derivative, it would be painful. We are going to establish several differentiation rules which you can use it and find the derivative of functions easily. We will begin with the simplest one.

 Derivative of Constant Function 常数函数导数

If $f(x) = c$, a constant function; then $f'(x) = 0$.
如果 $f(x) = c$ 是一个常数函数，那么，$f'(x) = 0$。

 Proof 证明

$$f'(x) = \lim_{h \to 0} \frac{f(x+h) - f(x)}{h} = \lim_{h \to 0} \frac{c - c}{h} = \lim_{h \to 0} \frac{0}{h} = 0.$$

 Power Rule 幂规律

If $f(x) = x^n$, then $f'(x) = nx^{n-1}$.
如果 $f(x) = x^n$，那么，$f'(x) = nx^{n-1}$。

 Proof 证明

$$f'(x) = \lim_{h \to 0} \frac{f(x+h) - f(x)}{h} = \lim_{h \to 0} \frac{(x+h)^n - x^n}{h} = \lim_{h \to 0} \frac{x^n + nx^{n-1}h + \frac{n(n-2)}{2}x^{n-2}h^2 + \cdots - x^n}{h}$$

$$= \lim_{h \to 0} \frac{nx^{n-1}h + \frac{n(n-2)}{2}x^{n-2}h^2 + \cdots}{n} = \lim_{h \to 0} \frac{h\left(nx^{n-1} + \frac{n(n-2)}{2}x^{n-2}h + \cdots\right)}{h} = nx^{n-1}.$$

Note 注意

We use the binomial theorem in the proof, so students may have question that what if n is not a positive integer. The binomial theorem is valid when n is not a positive integer, indeed; when n is a rational number or negative number, the expansion is an infinite series which we called Binomial Series.
We will deal with Binomial Series in Pure Math 3.

我们在证明中使用二项式定理,所以学生可能会有疑问,如果 n 不是正整数呢？二项式定理在 n 是非正整数时也是有效的.当 n 是有理数或负数时,我们将一个无穷级数的展开称为二项式级数.我们将在纯数 3 中讨论二项式级数.

📖 The Constant Multiple Rule　常数倍数法则

If c is a constant and $f(x)$ is a differentiable function, then $\frac{\mathrm{d}}{\mathrm{d}x}(c \cdot f(x)) = c \cdot \frac{\mathrm{d}}{\mathrm{d}x}(f(x)) = c \cdot f'(x)$.

如果 c 是常数,而且 $f(x)$ 是可微函数,那么,$\frac{\mathrm{d}}{\mathrm{d}x}(c \cdot f(x)) = c \cdot \frac{\mathrm{d}}{\mathrm{d}x}(f(x)) = c \cdot f'(x)$.

💻 Proof　证明

$$\frac{\mathrm{d}}{\mathrm{d}x}(c \cdot f(x)) = \lim_{h \to 0} \frac{cf(x+h) - cf(x)}{h} = c \lim_{h \to 0} \frac{f(x+h) - f(x)}{h} = c \cdot f'(x). \blacksquare$$

📖 The Sum and Difference Rules　函数和与差的求导法则

If $f(x)$ and $g(x)$ are both differentiable functions, then
$\frac{\mathrm{d}}{\mathrm{d}x}(f(x) \pm g(x)) = \frac{\mathrm{d}}{\mathrm{d}x}(f(x)) \pm \frac{\mathrm{d}}{\mathrm{d}x}(g(x)) = f'(x) \pm g'(x)$.

如果 $f(x)$ 和 $g(x)$ 都是可微函数,那么,$\frac{\mathrm{d}}{\mathrm{d}x}(f(x) \pm g(x)) = \frac{\mathrm{d}}{\mathrm{d}x}(f(x)) \pm \frac{\mathrm{d}}{\mathrm{d}x}(g(x)) = f'(x) \pm g'(x)$.

💻 Proof　证明

Let $F(x) = f(x) \pm g(x)$.

$$F'(x) = \lim_{h \to 0} \frac{F(x+h) - F(x)}{h} = \lim_{h \to 0} \frac{[f(x+h) \pm g(x+h)] - [f(x) \pm g(x)]}{h}$$

$$= \lim_{h \to 0} \frac{[f(x+h) - f(x)] \pm [g(x+h) - g(x)]}{h}$$

$$= \lim_{h \to 0} \frac{[f(x+h) - f(x)]}{h} \pm \lim_{h \to 0} \frac{[g(x+h) - g(x)]}{h} = f'(x) \pm g'(x). \blacksquare$$

🍃 Example 例 8.4

Find the derivative of the followings：

(a) x^9；　(b) $\frac{1}{x^5}$；　(c) $f(x) = 4\sqrt{x} + \frac{1}{x}$；　(d) $y = x^2 - 3x + 5$.

✋ Solution 解

(a) $\frac{\mathrm{d}}{\mathrm{d}x}(x^9) = 9x^8$.

(b) $\dfrac{d}{dx}\left(\dfrac{1}{x^5}\right) = \dfrac{d}{dx}(x^{-5}) = -5x^{-6}$.

(c) $f(x) = 4\sqrt{x} + \dfrac{1}{x} = 4x^{\frac{1}{2}} + x^{-1} \Rightarrow f' = 4 \cdot \dfrac{1}{2}x^{\frac{-1}{2}} - x^{-2} = 2x^{\frac{-1}{2}} - x^{-2} = \dfrac{2}{\sqrt{x}} - \dfrac{1}{x^2}$.

(d) $y' = \dfrac{d}{dx}(x^2) - \dfrac{d}{dx}(3x) + \dfrac{d}{dx}(5) = 2x - 3 + 0 = 2x - 3$.

Example 8.5

Find the derivative of $g(t) = \dfrac{3}{t^4} + \dfrac{1}{t^3} - \dfrac{5}{t^2} + 7t - 100$ with respect to t.

Solution

$\dfrac{d}{dt}g(t) = g'(t) = \dfrac{d}{dt}\left(\dfrac{3}{t^4}\right) + \dfrac{d}{dt}\left(\dfrac{1}{t^3}\right) + \dfrac{d}{dt}\left(\dfrac{-5}{t^2}\right) + \dfrac{d}{dt}(7t) + \dfrac{d}{dt}(-100) = -12t^{-5} - 3t^{-4} + 10t^{-3} + 7$.

Example 8.6

Find the derivative of the followings:

(a) $(3x - 2)(2x^2 - 3x + 5)$;   (b) $(x + 1)(2x + 2)(3x + 3)$.

Solution

(a) $(3x - 2)(2x^2 - 3x + 5) = 6x^3 - 9x^2 + 15x - 4x^2 + 6x - 10 = 6x^3 - 13x^2 + 21x - 10$.

$\dfrac{d}{dx}((3x - 2)(2x^2 - 3x + 5)) = \dfrac{d}{dx}(6x^3 - 13x^2 + 21x - 10) = 18x^2 - 26x + 21$.

(b) $(x + 1)(2x + 2)(3x + 3) = 6(x + 1)^3 = 6(x^3 + 3x^2 + 3x + 1)$.

$\dfrac{d}{dx}((x + 1)(2x + 2)(3x + 3)) = \dfrac{d}{dx}6(x^3 + 3x^2 + 3x + 1) = 6 \cdot \dfrac{d}{dx}(x^3 + 3x^2 + 3x + 1) = 6(3x^2 + 6x + 3)$.

Example 8.7

Suppose $f(x)$ is a polynomial satisfies $f(1) = 2$ and $3f(x) - xf'(x) - 3 = 0$, $\forall x$. Find $f(x)$.

Solution

Let $f(x) = a_n x^n + a_{n-1} x^{n-1} + \cdots + a_1 x + a_0$

$\Rightarrow 3(a_n x^n + a_{n-1} x^{n-1} + \cdots + a_1 x + a_0) - x[n \cdot a_n x^{n-1} + (n-1)a_{n-1} x^{n-2} + \cdots + a_1] - 3 = 0$

$\Rightarrow 3(a_n x^n + a_{n-1} x^{n-1} + \cdots + a_1 x + a_0) - [n \cdot a_n x^n + (n-1)a_{n-1} x^{n-1} + \cdots + a_1 x + 3] = 0$

$\Rightarrow (3a_n - na_n)x^n + \cdots = 0$.

Hence, $n = 3$, $f(x)$ is a cubic equation; so we can let $f(x) = ax^3 + bx^2 + cx + d$.

$3f(x) - xf'(x) - 3 = 0 \Rightarrow 3(ax^3 + bx^2 + cx + d) - x(3ax^2 + 2bx + c) - 3 = 0$

$\Rightarrow 3ax^3 - 3ax^3 + 3bx^2 - 2bx^2 + 3cx - cx + 3d - 3 = 0$

$\Rightarrow bx^2 + 2cx + 3d - 3 = 0, \forall x \Rightarrow b = 0, c = 0, d = 1$.

So $f(x) = ax^3 + 1$ and $f(1) = 2 \Rightarrow a + 1 = 2 \Rightarrow a = 1$.
Therefore, $f(x) = x^3 + 1$.

Example 例 8.8

Find the gradient of the tangent and normal lines to the curve $y = x^2 + 4x - 5 + \dfrac{2}{x}$ at the point $(1, 2)$.

Solution 解

The gradient of the tangent to the curve at $(1, 2)$ is just equal to $y'|_{x=1}$.

$y' = 2x + 4 - \dfrac{2}{x^2} \Rightarrow y'|_{x=1} = 2 + 4 - 2 = 4$.

So the gradient of the tangent is 4, and the gradient of the normal is $\dfrac{-1}{4}$.

Example 例 8.9

Find the equations of the tangent and normal lines to the curve $y = x^2 + 3x - 5$ at the point $(1, -1)$.

Solution 解

$y' = 2x + 3$, $y'|_{x=1} = 2 + 3 = 5$.

So the equation of the tangent is $\dfrac{y+1}{x-1} = 5 \Rightarrow y = 5x - 6$, and the equation of the normal is $\dfrac{y+1}{x-1} = \dfrac{-1}{5} \Rightarrow$

$y = \dfrac{-1}{5}x - \dfrac{4}{5}$.

Example 例 8.10

Given $f(x) = \dfrac{5}{x} + 2$ and $A(1, 7)$, the tangent line to the curve at A intersects the x-axis at P and y-axis at Q.
Find the distance PQ and the angle that this tangent line makes with the x-axis.

Solution 解

$f' = \dfrac{-5}{x^2} \Rightarrow f'(1) = -5$. So the equation of the tangent line at $A(1, 7)$ is $\dfrac{y-7}{x-1} = -5 \Rightarrow y = -5x + 12$.

Hence, $P\left(\dfrac{12}{5}, 0\right)$ and $Q(0, 12)$. $|PQ| = \sqrt{\left(\dfrac{12}{5}\right)^2 + 12^2} = \dfrac{12}{5}\sqrt{26}$.

$\angle QPO = \tan^{-1}\left(\dfrac{12}{\frac{12}{5}}\right) = 78.7°$ corrected to one decimal place.

Chapter 8 Differentiation

8.7 The Chain Rule and Higher Order Derivative
链式法则与高阶导数

By now we only know $\dfrac{d}{dx}x^n = n \cdot x^{n-1}$ and some other basic differential rules. If we want to find the derivative of $(ax+b)^{100}$, then we have to expand the expression and use the rules we have known to find the derivative which makes the calculation painful. Hence, we need a more powerful technique to make life easier; it is called the Chain Rule.

If $f(x)$ and $g(x)$ are both differentiable functions, that is f' and g' exist. Then the composite function $F = f(g(x))$ is a differentiable function and $F' = f'(g(x)) \times g'(x)$.

In Leibniz's notation, if $y = f(u)$ and $u = g(x)$; then $\dfrac{dy}{dx} = \dfrac{dy}{du} \cdot \dfrac{du}{dx}$.

📝 The Chain Rule 链式法则

If $f(x)$ and $g(x)$ are both differentiable functions. Then $f(g(x))$ is a differentiable function and
$$\dfrac{d}{dx}(f(g(x))) = f'(g(x)) \cdot g'(x).$$

如果 $f(x)$ 和 $g(x)$ 都是可微函数，那么，$f(g(x))$ 也是可微函数，并且 $\dfrac{d}{dx}(f(g(x))) = f'(g(x)) \cdot g'(x)$.

Suppose $f(x)$ and $f'(x)$ are differentiable functions. Then we can find $\dfrac{d}{dx}(f'(x))$ which is denoted by $f''(x)$ and this is called the second derivative of $f(x)$.

Using Leibniz's notation, $\dfrac{d}{dx}\left(\dfrac{dy}{dx}\right) = \dfrac{d^2y}{dx^2}$.

So $\dfrac{d^3y}{dx^3} = \dfrac{d}{dx}\left(\dfrac{d^2y}{dx^2}\right) = y'''$.

In general, $\dfrac{d^n y}{dx^n} = \dfrac{d}{dx}\left(\dfrac{d^{n-1}y}{dx^{n-1}}\right)$; when n gets larger, we write the nth order derivative as $f^{(n)}(x)$.

For example, $\dfrac{d^2y}{dx^2} = y'' = y^{(2)}(x)$ and $\dfrac{d^3y}{dx^3} = y'''(x) = y^{(3)}(x)$ etc.

📖 Example 例 8.11

Differentiate the followings:

(a) $(x+2)^{100}$;   (b) $(3x-5)^7$;   (c) $\dfrac{3}{(2x+1)^5}$;   (d) $\sqrt{x^3 - 2x + 1}$.

Solution 解

(a) $\dfrac{d}{dx}(x+2)^{100} = 100(x+2)^{99} \cdot \dfrac{d}{dx}(x+2) = 100(x+2)^{99} \cdot 1 = 100(x+2)^{99}$.

Or

let $u = x+2$, $y = u^{100} \Rightarrow \dfrac{du}{dx} = 1$ and $\dfrac{dy}{du} = 100u^{99}$, so $\dfrac{dy}{dx} = \dfrac{dy}{du} \cdot \dfrac{du}{dx} = 100u^{99} \cdot 1 = 100(x+2)^{99}$.

(b) $\dfrac{d}{dx}(3x-5)^{7} = 7(3x-5)^{6} \cdot \dfrac{d}{dx}(3x-5) = 7(3x-5)^{6} \cdot 3 = 21(3x-5)^{6}$.

Or

let $u = 3x-5$, $y = u^{7} \Rightarrow \dfrac{du}{dx} = 3$ and $\dfrac{dy}{du} = 7u^{6}$, so $\dfrac{dy}{dx} = \dfrac{dy}{du} \cdot \dfrac{du}{dx} = 7u^{6} \cdot 3 = 21(3x-5)^{6}$.

(c) $\dfrac{d}{dx}\left(\dfrac{3}{(2x+1)^{5}}\right) = \dfrac{d}{dx}(3(2x+1)^{-5}) = 3 \cdot -5(2x+1)^{-6} \cdot \dfrac{d}{dx}(2x+1) = 3 \cdot -5(2x+1)^{-6} \cdot 2 = -30(2x+1)^{-6}$.

Or

let $u = 2x+1$, $y = 3u^{-5} \Rightarrow \dfrac{du}{dx} = 2$ and $\dfrac{dy}{du} = -15u^{-6}$, so $\dfrac{dy}{dx} = \dfrac{dy}{du} \cdot \dfrac{du}{dx} = -15u^{-6} \cdot 2 = -30(2x+1)^{-6}$.

(d) $\dfrac{d}{dx}\sqrt{x^{3}-2x+1} = \dfrac{d}{dx}(x^{3}-2x+1)^{\frac{1}{2}} = \dfrac{1}{2}(x^{3}-2x+1)^{-\frac{1}{2}} \cdot \dfrac{d}{dx}(x^{3}-2x+1) = \dfrac{1}{2}(x^{3}-2x+1)^{-\frac{1}{2}} \cdot (3x^{2}-2)$.

Or

let $u = x^{3}-2x+1$, $y = u^{\frac{1}{2}} \Rightarrow \dfrac{du}{dx} = 3x^{2}-2$ and $\dfrac{dy}{du} = \dfrac{1}{2}u^{-\frac{1}{2}}$, so

$\dfrac{dy}{dx} = \dfrac{dy}{du} \cdot \dfrac{du}{dx} = \dfrac{1}{2}u^{-\frac{1}{2}} \cdot (3x^{2}-2) = \dfrac{1}{2}(x^{3}-2x+1)^{-\frac{1}{2}}(3x^{2}-2)$. ■

Example 例 8.12

Given $f(x) = (3x-1)^{2}$, the tangent lines at $A(1, 4)$ and $B(-1, 16)$ meet at P. Find the coordinate of P.

Solution 解

$f'(x) = 2(3x-1) \cdot 3 = 6(3x-1)$.

$f'(1) = 12$ and $f'(-1) = -24$. So the equations for the tangent lines are $\dfrac{y-4}{x-1} = 12$ and $\dfrac{y-16}{x+1} = -24$.

$y = 12x - 8$ and $y = -24x - 8$.

Solve the simultaneous equations, we get $x = 0$ and $y = -8$, so $P(0, -8)$. ■

Example 例 8.13

The curve passes through the point $A(1, 2)$ and has equation $y = \dfrac{4}{x+1}$. The normal line to the curve at point A meets the curve at point P. Find the coordinate of P and the perpendicular bisector of AP.

Solution

$y = \dfrac{4}{x+1} = 4(x+1)^{-1} \Rightarrow y' = -4(x+1)^{-2}$, the gradient of normal line is $\left.\dfrac{-1}{y'}\right|_{x=1} = \dfrac{-1}{-1} = 1$.

The equation of the normal is $\dfrac{y-2}{x-1} = 1 \Rightarrow y = x+1$. To find the intersection point and the curve, we equate

$\dfrac{4}{x+1} = x+1 \Rightarrow x^2 + 2x - 3 = 0 \Rightarrow (x-1)(x+3) = 0 \Rightarrow x = 1, -3$.

Therefore, $P(-3, -2)$.

Let M be the midpoint of AP, so $M\left(\dfrac{1+(-3)}{2}, \dfrac{2+(-2)}{2}\right) \Rightarrow M(-1, 0)$.

The equation of the perpendicular bisector would be $\dfrac{y-0}{x-(-1)} = -1 \Rightarrow y = -x - 1$. ∎

Example 8.14

Given $f(x) = \dfrac{3}{(5x+1)^4}$, find the first and second derivative of $f(x)$.

Solution

$f(x) = \dfrac{3}{(5x+1)^4} = 3(5x+1)^{-4}$.

$f'(x) = 3(-4)(5x+1)^{-5} \cdot 5 = -60(5x+1)^{-5}$.

$f''(x) = (-60)(-5)(5x+1)^{-6} \cdot 5 = 1\,500(5x+1)^{-6}$. ∎

Example 8.15

A curve has equation $y = 3x + \dfrac{1}{x} + 4$, and the points $A\left(\dfrac{1}{2}, 7\dfrac{1}{2}\right)$ and $B\left(4, 16\dfrac{1}{4}\right)$ on lie on the curve.

There are two points C and D also lie on the curve, and their tangent to the curve at C and D is parallel to AB. Find the perpendicular bisector of CD.

Solution

The gradient of $AB = \dfrac{16\dfrac{1}{4} - 7\dfrac{1}{2}}{4 - \dfrac{1}{2}} = 2\dfrac{1}{2}$.

$y' = 3 - \dfrac{1}{x^2} = 2\dfrac{1}{2} \Rightarrow \dfrac{1}{x^2} = \dfrac{1}{2} \Rightarrow x = \pm\sqrt{2}$.

When $x = \sqrt{2} \Rightarrow y = 3\sqrt{2} + \dfrac{1}{\sqrt{2}} + 4 = 4 + \dfrac{7}{2}\sqrt{2}$.

When $x = -\sqrt{2} \Rightarrow y = 3(-\sqrt{2}) + \dfrac{1}{-\sqrt{2}} + 4 = 4 - \dfrac{7}{2}\sqrt{2}$.

The midpoint M of CD has coordinate $(0, 4)$.

Therefore, the perpendicular bisector has equation $\dfrac{y-4}{x-0} = -\dfrac{2}{5} \Rightarrow y = -\dfrac{2}{5}x + 4$.

Example 8.16

Find $\dfrac{d}{dx}\left(\dfrac{2x+7}{2x-1}\right)^{10}$.

Solution

We do not know the differentiation rules for product of functions or quotient of functions at this moment. Therefore, we have to simplify the expression to something we know how to deal with.

$\dfrac{2x+7}{2x-1} = \dfrac{2x-1+8}{2x-1} = 1 + \dfrac{8}{2x-1}$.

$\dfrac{d}{dx}\left(\dfrac{2x+7}{2x-1}\right)^{10} = \dfrac{d}{dx}\left(\left(1+\dfrac{8}{2x-1}\right)^{10}\right) = 10\left(1+\dfrac{8}{2x-1}\right)^9 \cdot \dfrac{d}{dx}\left(1+\dfrac{8}{2x-1}\right)$

$= 10\left(1+\dfrac{8}{2x-1}\right)^9 \cdot \dfrac{d}{dx}\left(\dfrac{8}{2x-1}\right) = 10\left(1+\dfrac{8}{2x-1}\right)^9 \cdot \dfrac{d}{dx}(8(2x-1)^{-1})$

$= 10\left(1+\dfrac{8}{2x-1}\right)^9 \cdot (-8)(2x-1)^{-2} \cdot 2 = -160(2x-1)^{-2}\left(1+\dfrac{8}{2x-1}\right)^9$

$= -160(2x-1)^{-2}\left(\dfrac{2x+7}{2x-1}\right)^9 = \dfrac{-160(2x+7)^9}{(2x-1)^{11}}$.

Example 8.17

Given a function $f(x)$ satisfies $f(x^3 - 8) = 6x^2 - 4x + 11$. Find the value of $f'(0)$.

Solution

$\dfrac{d}{dx}[f(x^3-8)] = 6x^2 - 4x + 11 \Rightarrow 3x^2 \cdot f'(x^3-8) = 12x - 4$.

$x^3 - 8 = 0 \Rightarrow x = 2$.

$3(2)^2 \cdot f'(0) = 12(2) - 4 \Rightarrow 12f'(0) = 20 \Rightarrow f'(0) = \dfrac{5}{3}$.

Example 8.18

Differentiate $f(x) = \dfrac{3x^2 - 5x + 1}{x}$.

Solution

$f(x) = \dfrac{3x^2 - 5x + 1}{x} = 3x - 5 + \dfrac{1}{x}$, $f'(x) = 3 - \dfrac{1}{x^2}$.

Example 8.19

The line $2y - x = 5$ is a tangent line to the curve $y = 2x^2 - 3x + k$. Find the value of k.

Solution 解

$2y - x = 5 \Rightarrow y = \dfrac{1}{2}x + \dfrac{5}{2} \Rightarrow$ The gradient of the tangent line is $\dfrac{1}{2}$.

$y' = 4x - 3 = \dfrac{1}{2} \Rightarrow x = \dfrac{7}{8}$.

When $x = \dfrac{7}{8} \Rightarrow y = \dfrac{1}{2} \cdot \dfrac{7}{8} + \dfrac{5}{2} = \dfrac{47}{16}$.

$\dfrac{47}{16} = 2\left(\dfrac{7}{8}\right)^2 - 3\left(\dfrac{7}{8}\right) + k \Rightarrow k = \dfrac{129}{32} = 4\dfrac{1}{32}$. ∎

Example (例) 8.20

The line $2y - x = 5$ is a normal line to the curve $y = 2x^2 - 3x + k$. Find the value of k.

Solution 解

$2y - x = 5 \Rightarrow y = \dfrac{1}{2}x + \dfrac{5}{2} \Rightarrow$ The gradient of the tangent line is $\dfrac{-1}{\frac{1}{2}} = -2$.

$y' = 4x - 3 = -2 \Rightarrow x = \dfrac{1}{4}$.

When $x = \dfrac{1}{4} \Rightarrow y = \dfrac{1}{2} \cdot \dfrac{1}{4} + \dfrac{5}{2} = \dfrac{21}{8}$.

$\dfrac{21}{8} = 2\left(\dfrac{1}{4}\right)^2 - 3\left(\dfrac{1}{4}\right) + k \Rightarrow k = \dfrac{13}{4} = 3\dfrac{1}{4}$. ∎

Summary of Key Theories 核心定义总结

(1) If $f(x)$ is defined for all x near $x = a$, except possibly at $x = a$. If $f(x)$ approaches the value L as x approaches a, then we write $\lim\limits_{x \to a} f(x) = L$.

(2) If $f(x)$ is a polynomial, rational or trigonometric function, then $\lim\limits_{x \to a} f(x) = f(a)$ provided that $x = a$ is in the domain of $f(x)$.

(3) The derivative of a function $f(x)$ is another function denoted by $f'(x)$ (read "f prime"), where $f'(x) = \lim\limits_{h \to 0} \dfrac{f(x+h) - f(x)}{h}$ at all points x for which the limit exists.

(4) If $f'(x)$ exist at a point $x = x_0$, we say the function $f(x)$ is differentiable at $x = x_0$.

(5) If $f(x) = c$, a constant function; then $f'(x) = 0$.

(6) If $f(x) = x^n$, then $f'(x) = nx^{n-1}$.

(7) If c is a constant and $f(x)$ is a differentiable function; then $\dfrac{\mathrm{d}}{\mathrm{d}x}(c \cdot f(x)) = c \cdot \dfrac{\mathrm{d}}{\mathrm{d}x}(f(x)) = c \cdot f'(x)$.

(8) If $f(x)$ and $g(x)$ are both differentiable functions, then $\dfrac{d}{dx}(f(x) \pm g(x)) = \dfrac{d}{dx}(f(x)) \pm \dfrac{d}{dx}(g(x)) = f'(x) \pm g'(x)$.

(9) If $f(x)$ and $g(x)$ are both differentiable functions. Then $f(g(x))$ is a differentiable function and $\dfrac{d}{dx}(f(g(x))) = f'(g(x)) \cdot g'(x)$.

Applications of Differentiation

第 9 章 微 分 的 应 用

9.1 Increasing and Decreasing Function 递增与递减函数

Definition 定义

A function $f(x)$ is said to be increasing on an interval I, if $f(x_2) > f(x_1)$ whenever $x_2 > x_1$.

A function $f(x)$ is said to be decreasing on an interval I, if $f(x_2) < f(x_1)$ whenever $x_2 > x_1$.

A function $f(x)$ is said to be nonincreasing on an interval I, if $f(x_2) \leqslant f(x_1)$ whenever $x_2 > x_1$.

A function $f(x)$ is said to be nondecreasing on an interval I, if $f(x_2) \geqslant f(x_1)$ whenever $x_2 > x_1$.

如果 $f(x_2) > f(x_1)$，当 $x_2 > x_1$ 时，函数 $f(x)$ 在区间 I 递增.

如果 $f(x_2) < f(x_1)$，当 $x_2 > x_1$ 时，函数 $f(x)$ 在区间 I 递减.

如果 $f(x_2) \leqslant f(x_1)$，当 $x_2 > x_1$ 时，函数 $f(x)$ 在区间 I 非递增.

如果 $f(x_2) \geqslant f(x_1)$，当 $x_2 > x_1$ 时，函数 $f(x)$ 在区间 I 非递减.

Since we know the derivative represents the gradient of the tangent to the curve, we can use mean value theorem (we did not discuss in this course) to prove the following theorem.

Theorem 9.1 定理 9.1

Suppose a function $f(x)$ is continuous on an interval I and differentiable on all the points in I except the endpoints.

If $f'(x) > 0$ in I, then $f(x)$ is increasing on I.

If $f'(x) < 0$ in I, then $f(x)$ is decreasing on I.

If $f'(x) \leq 0$ in I, then $f(x)$ is nonincreasing on I.

If $f'(x) \geq 0$ in I, then $f(x)$ is nondecreasing on I.

假设一个函数 $f(x)$ 在区间 I 上是连续的，并且在区间 I 除端点外的所有点都是可微的：

如果在区间 I 上，$f'(x) > 0$，则 $f(x)$ 在 I 上递增.

如果在区间 I 上，$f'(x) < 0$，则 $f(x)$ 在 I 上递减.

如果在区间 I 上，$f'(x) \leq 0$，则 $f(x)$ 在 I 上非递增.

如果在区间 I 上，$f'(x) \geq 0$，则 $f(x)$ 在 I 上非递减.

Proof 证明

Let x_1 and x_2 be point in the interval I with $x_2 > x_1$.

From Mean-Value Theorem, we know $\dfrac{f(x_2) - f(x_1)}{x_2 - x_1} = f'(c)$, for some $c \in I$.

$f(x_2) - f(x_1) = (x_2 - x_1)f'(c)$. Since $(x_2 - x_1) > 0$, according to the definition we have these four results.

Example 例 9.1

Find the intervals the function $f(x)$ is increasing or decreasing.

(a) $f(x) = x^3 + 7x^2 - 5x - 9$； (b) $f(x) = -4x^3 + 15x^2 + 18x + 100$； (c) $f(x) = 3x^4 - 16x^3 + 24x^2 - 9$.

Solution 解

(a) $f' = 3x^2 + 14x - 5 = (x + 5)(3x - 1)$ which is a parabola open upwards.

So $f' > 0$ when $x \in (-\infty, -5) \cup \left(\dfrac{1}{3}, \infty\right)$, and $f' < 0$ when $x \in \left(-5, \dfrac{1}{3}\right)$.

Therefore, the function is increasing on $(-\infty, -5) \cup \left(\dfrac{1}{3}, \infty\right)$, and the function is decreasing on $\left(-5, \dfrac{1}{3}\right)$.

(b) $f' = -12x^2 + 30x + 18 = 6(-2x^2 + 5x + 3) = 6(-x + 3)(2x + 1)$ which is a parabola open downwards.

So $f' > 0$ when $x \in \left(-\dfrac{1}{2}, 3\right)$, and $f' < 0$ when $x \in \left(-\infty, -\dfrac{1}{2}\right) \cup (3, \infty)$.

Therefore, the function is increasing on $\left(-\dfrac{1}{2}, 3\right)$, and the function is decreasing on $\left(-\infty, -\dfrac{1}{2}\right) \cup (3, \infty)$.

(c) $f' = 12x^3 - 48x^2 + 48x = 12x(x^2 - 4x + 4) = 12x(x - 2)^2$.

So $f' > 0$ when $x > 0$ and $x \neq 2$, $f' < 0$ when $x < 0$.

Therefore, the function is increasing on $(0, 2) \cup (2, \infty)$, and the function is decreasing on $(-\infty, 0)$.

Chapter 9 Applications of Differentiation

Example 9.2

A function $f(x)$ is defined as $f(x) = \dfrac{6}{1-3x}$ for $x > \dfrac{1}{3}$. Prove that it is an increasing function.

Solution

$f(x) = \dfrac{6}{1-3x} = 6(1-3x)^{-1} \Rightarrow f'(x) = -6(1-3x)^{-2} \cdot (-3) = \dfrac{18}{(1-3x)^2}.$

The derivative is always positive when $x > \dfrac{1}{3}$, therefore $f(x)$ is always increase in the domain of $f(x)$. Hence, it is an increasing function.

Example 9.3

(a) Prove $f(x) = x^3 - 2x^2 + 7x - 17$ is an increasing function.
(b) Prove $g(x) = -2x^3 + x^2 - 5x + 19$ is a decreasing function.

Solution

(a) $f' = 3x^2 - 4x + 7 = 3\left(x^2 - \dfrac{4}{3}x\right) + 7 = 3\left(x^2 - \dfrac{4}{3}x + \dfrac{4}{9} - \dfrac{4}{9}\right) + 7$

$= 3\left(x^2 - \dfrac{4}{3}x + \dfrac{4}{9}\right) - \dfrac{4}{3} + 7 = 3\left(x - \dfrac{2}{3}\right)^2 + \dfrac{17}{3} \geqslant \dfrac{17}{3} > 0.$

The derivative is always positive regardless the values of x, therefore the function is an increasing function.

(b) $g' = -6x^2 + 2x - 5 = -6\left(x^2 - \dfrac{1}{3}x\right) - 5 = -6\left(x^2 - \dfrac{1}{3}x + \dfrac{1}{36} - \dfrac{1}{36}\right) - 5$

$= -6\left(x^2 - \dfrac{1}{3}x + \dfrac{1}{36}\right) + \dfrac{1}{6} - 5 = -6\left(x - \dfrac{1}{6}\right)^2 - \dfrac{29}{6} \leqslant -\dfrac{29}{6} < 0.$

The derivative is always negative regardless the values of x, therefore the function is a decreasing function.

Example 9.4

Given that a function $f(x) = \dfrac{1}{3}x^3 + (m-5)x^2 + 2(3m-19)x - 20$ is an increasing function $\forall x \in \mathbf{R}$. Find the value of m.

Solution

$f' = x^2 + 2(m-5)x + 2(3m-19) > 0$

$\Rightarrow D = 4(m-5)^2 - 4 \cdot 2(3m-19) < 0 \Rightarrow 4(m^2 - 10m + 25) - 8(3m-19) < 0$

$\Rightarrow m^2 - 16m + 63 < 0 \Rightarrow (m-7)(m-9) < 0.$

Hence, $7 < m < 9$.

Example 9.5

Find the interval of increasing or decreasing of $f(x) = \sqrt{1-x^2}$, for $-1 < x < 1$.

 **解**

$$f(x) = \sqrt{1-x^2} = (1-x^2)^{\frac{1}{2}} \Rightarrow f' = \frac{1}{2}(1-x^2)^{\frac{-1}{2}}(-2x) = \frac{-x}{\sqrt{1-x^2}}.$$

The denominator $\sqrt{1-x^2}$ is always positive for $-1 < x < 1$, so the numerator $-x$ determines the sign of $f'(x)$.

When $-1 < x < 0$, $f' > 0 \Rightarrow f(x)$ is increasing.

When $\quad 0 < x < 1$, $f' < 0 \Rightarrow f(x)$ is decreasing.

The function is increasing on the interval $(-1, 0)$, decreasing on the interval $(0, 1)$. ∎

9.2 Maximum/Minimum and Stationary Points
最大值/最小值及驻点

One of the most important applications in differential calculus is finding the maximum or minimum values which are called optimization problems. We will begin with some definitions of maximum and minimum points.

> **Definition 定义**
>
> A function $f(x)$ is said to have an absolute maximum value (or global maximum) $f(x_0)$ at the point x_0 in the domain of $f(x)$ if $f(x) \leq f(x_0)$ for all x in the domain of $f(x)$.
> 一个函数$f(x)$被认为在定义域的点x_0存在绝对最大值(全局最大值)$f(x_0)$,如果$f(x) \leq f(x_0)$对所有位于定义域的值x成立.
>
> A function $f(x)$ is said to have an absolute minimum value (or global minimum) $f(x_0)$ at the point x_0 in the domain of $f(x)$ if $f(x) \geq f(x_0)$ for all x in the domain of $f(x)$.
> 一个函数$f(x)$被认为在定义域的点x_0存在绝对最小值(全局最小值)$f(x_0)$,如果$f(x) \geq f(x_0)$对所有位于定义域的值x成立.
>
> A function $f(x)$ is said to have a local maximum value (or relative maximum) $f(x_0)$ at the point x_0 in the domain of $f(x)$ if $\exists h > 0$ such that $f(x) \leq f(x_0)$ whenever x is in the domain of $f(x)$ and $|x - x_0| < h$.
> 一个函数$f(x)$被认为在定义域的点x_0存在局部最大值(相对最大值)$f(x_0)$,如果$\exists h > 0$时$f(x) \leq f(x_0)$对所有位于定义域且满足$|x - x_0| < h$的x成立.
>
> A function $f(x)$ is said to have a local minimum value (or relative minimum) $f(x_0)$ at the point x_0 in the domain of $f(x)$ if $\exists h > 0$ such that $f(x) \geq f(x_0)$ whenever x is in the domain of $f(x)$ and $|x - x_0| < h$.
> 一个函数$f(x)$被认为在定义域的点x_0存在局部最小值(相对最小值)$f(x_0)$,如果$\exists h > 0$时$f(x) \geq f(x_0)$对所有位于定义域且满足$|x - x_0| < h$的x成立.

The following figures show simply concepts of the local and absolute extreme values (Fig. 9.1).

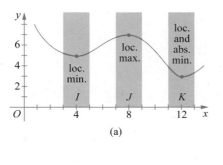
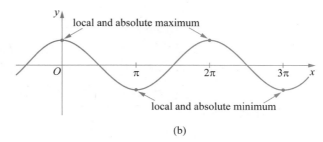
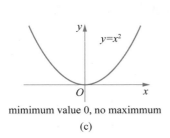
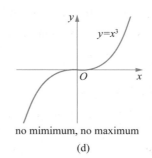

Fig. 9.1

Note 注意

In CAIE syllabus, they did not discuss both global and local extrema; they just simply use the word maximum or minimum. In the question, it is usually the local max. or min..

CAIE 大纲未涉及对全局和局部极值的讨论,只是简化为最大值或最小值,在题目里通常是指局部最大值或最小值.

Definition 定义

A stationary point of a function $f(x)$ is a point $(a, f(a))$ such that $f'(a) = 0$. The point $(a, f(a))$ could be a local maximum, local minimum or neither.

驻点指的是函数 $f(x)$ 上的点 $(a, f(a))$,满足 $f'(a) = 0$. 点 $(a, f(a))$ 可以是局部最大值、局部最小值,或者两者都不是.

When we find the stationary points, how can we identify whether they are maximum, minimum or neither? We therefore need a test that will tell us whether or not the point is a local maximum or minimum or neither. This test is known as the first derivative test.

The First Derivative Test 一阶导数检测

Suppose $(a, f(a))$ is a stationary point for the function $f(x)$.

(a) If f' changes form positive to negative at $x = a$, then $(a, f(a))$ is a local maximum.

(b) If f' changes form negative to positive at $x = a$, then $(a, f(a))$ is a local minimum.

(c) If f' does not change signs at $x = a$ (positive to positive or negative to negative), then $(a, f(a))$ is neither a local maximum nor a local minimum.

> 假设 $(a, f(a))$ 是函数 $f(x)$ 的驻点.
> (a) 当 $x = a$, f' 从正值变成负值,则 $(a, f(a))$ 是局部最大值.
> (b) 当 $x = a$, f' 从负值变成正值,则 $(a, f(a))$ 是局部最小值.
> (c) 当 $x = a$, f' 不变更符号(从正值到正值,从负值到负值),则 $(a, f(a))$ 既不是局部最大值,也不是局部最小值.

The local maximum or minimum point is sometimes called the turning point.

If a stationary point is neither maximum nor minimum, it may be an inflection point (inflexion point), we will give a formal definition later on.

Example 例 9.6

Find and classify all stationary points of $f(x) = x^3 - 3x^2 - 9x - 12$.

Solution 解

$f' = 3x^2 - 6x - 9 = 3(x^2 - 2x - 3) = 3(x-3)(x+1)$.

We can construct a "sign table" as the following (Tab. 9.1):

Tab. 9.1

x		-1		3	
f'	$+$	0	$-$	0	$+$

By the first derivative test, $(-1, f(-1)) \equiv (-1, -7)$ is maximum, $(3, f(3)) \equiv (3, -39)$ is minimum.

Example 例 9.7

Find the stationary points of $f(x) = \dfrac{x^2 - 2x + 5}{x}$ and determine the nature of the stationary points.

Solution 解

$f(x) = \dfrac{x^2 - 2x + 5}{x} = x - 2 + \dfrac{5}{x} \Rightarrow f' = 1 - \dfrac{5}{x^2} = 0 \Rightarrow x^2 = 5 \Rightarrow x = \pm\sqrt{5}$.

We can construct a "sign table" as the following (Tab. 9.2):

Tab. 9.2

x		$-\sqrt{5}$		0		$\sqrt{5}$	
f'	$+$	0	$-$	undefined	$-$	0	$+$

By the first derivative test, $(-\sqrt{5}, f(-\sqrt{5})) \equiv (-\sqrt{5}, -2\sqrt{5} - 2)$ is maximum, $(\sqrt{5}, f(\sqrt{5})) \equiv (\sqrt{5}, 2\sqrt{5} - 2)$ is minimum.

Similar to the first derivative, the second derivative of a function also provides useful information about the behavior of the function and the shape of the graph. The second derivative determines whether the graph is bending upwards or downwards.

Definition 定义

The function $f(x)$ is said to be concave up on an interval I, if $f'(x)$ is an increasing function on I. Similarly, a function $f(x)$ is concave down on an interval I, if $f'(x)$ is a decreasing function on I. In other words, if the graph of $f(x)$ lies above all of its tangents on I, then it is concave upwards on I. If the graph of $f(x)$ lies below all of its tangents on I, then it is concave downwards on I.

如果函数 $f'(x)$ 在区间 I 是递增函数,那么,函数 $f(x)$ 被认为在区间 I 上凹.
同理,如果函数 $f'(x)$ 在区间 I 是递减函数,那么,函数 $f(x)$ 被认为在区间 I 下凹.
换句话说,如果函数 $f(x)$ 的图像位于 I 的所有切线之上,那么,它在区间 I 上凹;如果函数 $f(x)$ 的图像位于 I 的所有切线之下,那么,它在区间 I 下凹.

The second derivative can also help us to determine the nature of the stationary point. The test we are using is called the second derivative test or concavity test.

The Second Derivative Test (Concavity Test) 二阶导数检测(凹度检测)

Suppose $(a, f(a))$ is a stationary point for the function $f(x)$.
(a) If $f''(a) < 0$, then $(a, f(a))$ is a local maximum.
(b) If $f''(a) > 0$, then $(a, f(a))$ is a local minimum.
(c) If $f'' = 0$, then no conclusion; it may be an inflection point, but not guaranteed.

假设 $(a, f(a))$ 是函数 $f(x)$ 的驻点.
(a) $f''(a) < 0$,则 $(a, f(a))$ 是局部最大值.
(b) $f''(a) > 0$,则 $(a, f(a))$ 是局部最小值.
(c) $f'' = 0$,则没有结论;它可能是一个拐点,但是不能确定.

Definition 定义

A point P on a curve $y = f(x)$ is an inflection point, if $f(x)$ is continuous and the curve changes concavity at P (either concave upwards to downwards or vice versa).

曲线 $y = f(x)$ 上的一个点 P 是拐点,如果 $f(x)$ 是连续的,并且曲线的凹处在 P 点变化(凹处由上转下,反之亦然).

For polynomial function, the second derivative test is a better one for determining the nature of the stationary points.

Example 例 9.8

Find the stationary points of $f(x) = x^3 - 3x^2 - 24x + 5$, and determining their natures.

Solution 解

$f' = 3x^2 - 6x - 24 = 3(x^2 - 2x - 8) = 3(x - 4)(x + 2) = 0 \Rightarrow x = -2, 4.$

$f'' = 6x - 6.$
$f''(-2) = -18 < 0 \Rightarrow (-2, 33)$ is maximum.
$f''(4) = 18 > 0 \Rightarrow (4, -75)$ is minimum.

Example 例 9.9

Use derivative to find the vertex of a parabola $y = ax^2 + bx + c$.

Solution 解

We know the vertex of the parabola is the maximum or minimum point for a parabola. Therefore, the vertex is a stationary point.

$y' = 2ax + b = 0 \Rightarrow x = \dfrac{-b}{2a}.$

The vertex of the parabola is $\left(\dfrac{-b}{2a}, \dfrac{4ac - b^2}{4a}\right).$

In general, for a polynomial with degree n; we have n roots and $n - 1$ stationary points.

Example 例 9.10

The equation $x^3 - 3x^2 - 24x - k = 0$ has three distinct real roots. Find the values of k.

Solution 解

Let $f(x) = x^3 - 3x^2 - 24x - k$, we first find the stationary points of the function.
$f' = 3x^2 - 6x - 24 = 0 \Rightarrow 3(x^2 - 2x - 8) = 0 \Rightarrow (x - 4)(x + 2) = 0.$
The stationary points are $(4, f(4))$ and $(-2, f(-2))$.
To determine the nature of these stationary points, we use the second derivative test.
$f'' = 6x - 6, f''(4) = 18 > 0 \Rightarrow (4, f(4))$ is minimum, and $f''(-2) = -18 < 0 \Rightarrow (-2, f(-2))$ is maximum.
In order to have three distinct roots, the stationary values $f(4) < 0$ and $f(-2) > 0$.
$f(4) = -80 - k < 0 \Rightarrow k > -80$ and $f(-2) = 28 - k > 0 \Rightarrow k < 28.$
Therefore, when $-80 < k < 28$, we have three distinct real roots.

Example 例 9.11

Given $f(x) = x^4 - 4n^3 x + 12 > 0$, $\forall x \in \mathbf{R}$. Find the range of n.

Solution 解

$f' = 4x^3 - 4n^3 = 4(x - n)(x^2 + nx + n^2) = 4(x - n)\left(\left(x + \dfrac{n}{2}\right)^2 + \dfrac{3}{4}n^2\right).$

It is clear that the stationary point is $(n, f(n))$.
$f'' = 12x^2 > 0$ if $x \neq 0$. Hence, the stationary point is a minimum.
$f(n) = n^4 - 4n^4 + 12 > 0 \Rightarrow n^4 < 4 \Rightarrow (n^2)^2 - 4 < 0 \Rightarrow -2 < n^2 < 2 \Rightarrow n^2 < 2 \Rightarrow n^2 - 2 < 0 \Rightarrow -\sqrt{2} < n < \sqrt{2}.$

Chapter 9　Applications of Differentiation　第九章　微分的应用

9.3　Related Rate　相关变化率

When two or more quantities that change with time are linked by an equation, then the equation can be differentiated with respect to time to produce an equation involves the rate of change of the quantities.

For example, if $x = h(t)$ and $y = g(t)$ are functions of time; however if we do not know the quantities x and y are related to time t, but we only know $y = f(x)$.

By the Chain Rule, we know $\dfrac{dy}{dx} = f'(x) = \dfrac{dy}{dt} \cdot \dfrac{dt}{dx} \Rightarrow \dfrac{dy}{dt} = \dfrac{dy}{dx} \cdot \dfrac{dx}{dt}$.

In a related rate problem, the idea is to compute the rate of change of one quantity in terms of the rate of change of others.

The strategy for solving related rate problems is:

(1) Construct a diagram is necessary.
(2) Collect all the given information or data.
(3) Create an equation that related the variables.
(4) Differentiate both sides of the equation with respect to time t.
(5) Substitute the given information to find the appropriate unknown rate.

Example 例 9.12

Given $y = t^2 - t + \dfrac{1}{t}$, find the rate of change of y when $t = 2$.

Solution 解

$y = t^2 - t + \dfrac{1}{t} \Rightarrow \dfrac{dy}{dt} = 2t - 1 - \dfrac{1}{t^2}$.

when $t = 2 \Rightarrow \dfrac{dy}{dt} = 4 - 1 - \dfrac{1}{4} = 2\dfrac{3}{4}$.

Example 例 9.13

Air is being pumped into a spherical balloon such that the volume is increasing at a rate of 150 cm³/s. How fast is the radius of the balloon increasing when the radius is 10 cm.

Solution 解

In this question, we are given $\dfrac{dV}{dt} = 150$, ask for $\dfrac{dr}{dt}$ when $r = 10$.

So we need to find a relationship between the volume V and radius r.

$V = \dfrac{4}{3}\pi r^3 \Rightarrow \dfrac{dV}{dt} = \dfrac{4}{3}\pi \cdot 3r^2 \cdot \dfrac{dr}{dt} \Rightarrow 150 = 4\pi(10)^2 \cdot \dfrac{dr}{dt} \Rightarrow \dfrac{dr}{dt} = \dfrac{150}{400\pi} = \dfrac{3}{8\pi}$ (cm/s).

Example 例 <9.14>

A point $P(x, y)$ is moving along the curve $y = 3x^2 + \dfrac{1}{5}x^{\frac{3}{2}}$ is such a way that the rate of change of y is constant. Find the value of x coordinate of the point P such that the rate of change of x is equal to one-third of the rate of change of y.

Solution 解

We are given $\dfrac{dy}{dt}$ is constant, ask for $\dfrac{dx}{dt}$.

$\dfrac{dy}{dt} = \left(6x + \dfrac{1}{5} \cdot \dfrac{3}{2}x^{\frac{1}{2}}\right) \cdot \dfrac{dx}{dt}$.

From the question, we know $\dfrac{dx}{dt} = \dfrac{1}{3}\dfrac{dy}{dt} \Rightarrow \dfrac{dy}{dt} = 3\dfrac{dx}{dt}$

$\Rightarrow 6x + \dfrac{1}{5} \cdot \dfrac{3}{2}x^{\frac{1}{2}} = 3 \Rightarrow 6(x^{\frac{1}{2}})^2 + \dfrac{3}{10}x^{\frac{1}{2}} - 3 = 0 \Rightarrow 60(x^{\frac{1}{2}})^2 + 3x^{\frac{1}{2}} - 30 = 0$

$\Rightarrow x^{\frac{1}{2}} = \dfrac{-3 \pm \sqrt{3^2 + 4 \cdot 60 \cdot 30}}{120} = \dfrac{-3 \pm \sqrt{7209}}{120}$, reject $\dfrac{-3 - \sqrt{7209}}{120}$ since it is negative

$\Rightarrow x = \left(\dfrac{-3 + \sqrt{7209}}{120}\right)^2 \approx 0.466$ corrected to three significant figures. ∎

Example 例 <9.15>

A point $P(x, y)$ is moving along the curve $y = 4x - \dfrac{6}{x}$ is such a way that the x-coordinate is decreasing at a constant rate of 0.1 units per second. Find the rate of change of the y-coordinate of the point P when $x = 2$.

Solution 解

It is given $\dfrac{dx}{dt} = -0.1$, ask $\dfrac{dy}{dt}$ when $x = 2$.

$y = 4x - \dfrac{6}{x} \Rightarrow \dfrac{dy}{dt} = \left(4 + \dfrac{6}{x^2}\right) \cdot \dfrac{dx}{dt}$.

When $x = 2 \Rightarrow \dfrac{dy}{dt} = \left(4 + \dfrac{6}{2^2}\right) \cdot (-0.1) = -0.55$.

So y-coordinate is decreasing at 0.55 units per second. ∎

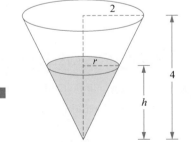

Fig. 9.2

Example 例 <9.16>

A leaky water tank is in the shape of an inverted right circular cone with base radius 2 m and height 4 m. If the water in the tank is leaking out at a rate of 1 m³/min, how fast is the water level in the tank dropping when the water in the tank is 3 m deep (Fig. 9.2)?

Solution 解

Let r and h be the surface radius and depth of water in the tank at time t.

We have been given $\frac{dV}{dt} = -1$ ask for $\frac{dh}{dt}$ when $h = 3$.

The volume V of water at the tank at time t is $V = \frac{1}{3}\pi r^2 h$.

Since we have two unknown variables r and h in the equation; therefore, we can not differentiate directly. We have to find an equation that relates V and h only.

Using similar triangle, we can find a relationship between r and h.

$\frac{r}{2} = \frac{h}{4} \Rightarrow r = \frac{h}{2} \Rightarrow V = \frac{1}{12}\pi h^3$.

So $\frac{dV}{dt} = \frac{1}{4}\pi h^2 \cdot \frac{dh}{dt} \Rightarrow -1 = \frac{1}{4}\pi \cdot 3^2 \cdot \frac{dh}{dt} \Rightarrow \frac{dh}{dt} = \frac{-4}{9\pi}$ (m/min). ∎

Example 例 9.17

The radius of a sphere is increasing at a rate of 1 cm/s. How fast is the volume and surface area increasing when the diameter is 10 cm?

Solution 解

Let the volume of the sphere be V, surface area be A and the radius be r.

We are given $\frac{dr}{dt} = 1$, ask for $\frac{dV}{dt}$ and $\frac{dA}{dt}$ when $r = 5$.

$V = \frac{4}{3}\pi r^3$ and $A = 4\pi r^2$ for sphere.

$\frac{dV}{dt} = 4\pi r^2 \cdot \frac{dr}{dt}$ and $\frac{dA}{dt} = 8\pi r \cdot \frac{dr}{dt}$.

So when $r = 5$: $\frac{dV}{dt} = 4\pi(5)^2 \cdot 1 = 100\pi (\text{cm}^3/\text{s})$, $\frac{dA}{dt} = 8\pi \cdot 5 \cdot 1 = 40\pi (\text{cm}^2/\text{s})$. ∎

9.4 Optimization Problems 优化问题

In this section, we are going to solve various word problems which would translate into mathematical terms require to find a maximum or minimum value of a function of a single variable. The problem involves finding a maximum or minimum value of a function is known as optimization problem. In optimization problem, we have an objective function which is the equation that we are interested in finding a maximum or minimum value of, and a constrain function which is usually given in the question. All we have to do is use the constraint function and substitute into the objective function to obtain a single variable function and then find the stationary points. The steps for solving optimization problems are as follows:

(1) Understand the question.
(2) Draw a diagram if possible.

(3) Find the objective function for the described problem.
(4) Substitute the constraint function into the objective function and then find the stationary points.

Example 例 9.18

Two real numbers x and y not necessary different, have a sum equal to 20. Find the maximum value of their product.

Solution 解

Let $P = xy$ be the objective function, and $x + y = 20$ is the constraint function.
$P(x) = xy = x(20 - x) = 20x - x^2$.
$P' = 20 - 2x = 0 \Rightarrow x = 10$.
$P'' = -2 \Rightarrow$ The stationary point is a maximum.
When $x = 10 \Rightarrow y = 10$, so the product is $P = 10 \times 10 = 100$.
The maximum value of their product is 100.

Example 例 9.19

A manufacturer wants to produce a cylindrical can that can hold 2 L of oil. Find the most economical shape of a cylindrical can that the manufacturer can produce (Fig. 9.3).

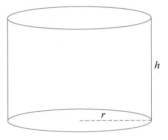

Fig. 9.3

Solution 解

We want the most economical shape of a cylindrical can that means we want to find the minimum surface area of the can.
Let the volume of the can be V and the surface area of the can be A.
Let the radius of the circle in the top and bottom of the can be r, the height of the can be h.
It is given that $V = \pi r^2 h = 2(\text{L}) = 2\,000(\text{mL})$ (constraint function).
Ask for maximum of $A = 2\pi r^2 + 2\pi rh$ (objective function).
$h = \dfrac{2\,000}{\pi r^2}$ from the constraint function.
$A(r) = A = 2\pi r^2 + 2\pi r \cdot \dfrac{2\,000}{\pi r^2} = 2\pi r^2 + \dfrac{4\,000}{r}$.
$\dfrac{dA}{dr} = 4\pi r - \dfrac{4\,000}{r^2} = 0 \Rightarrow 4\pi r^3 = 4\,000 \Rightarrow r = \dfrac{10}{(\pi)^{\frac{1}{3}}}$.
$\dfrac{d^2 A}{dr^2} = 4\pi + \dfrac{8\,000}{r^3}\bigg|_{r = \frac{10}{(\pi)^{\frac{1}{3}}}} > 0 \Rightarrow$ The stationary point is a minimum.

The manufacturer needs to design the cylindrical can for base circle with radius of $r = \dfrac{10}{(\pi)^{\frac{1}{3}}}$ cm, and the height of the can to be $h = \dfrac{20}{\pi^{\frac{1}{3}}}$ cm.

Chapter 9 Applications of Differentiation

Example 9.20

Find the point on the parabola $y^2 = 2x$ that is closest to the point $(1, 4)$.

Solution

Let (x, y) be the point that is closest to $(1, 4)$. So the distance D between the point and $(1, 4)$ is $D = \sqrt{(x-1)^2 + (y-4)^2}$ which is the objective function.

Since the point (x, y) lies on the parabola, so the constraint function is $y^2 = 2x$.

$$D(y) = \sqrt{\left(\frac{y^2}{2} - 1\right)^2 + (y-4)^2}.$$

If we can try to find $\dfrac{dD}{dy}$ directly, the calculation may be a disaster.

Think about this, the distance is always greater or equal to zero, so minimize D^2 is the same as minimize D.

We now, let $F = D^2 = \left(\dfrac{y^2}{2} - 1\right)^2 + (y-4)^2 \Rightarrow F' = 2\left(\dfrac{y^2}{2} - 1\right) \cdot y + 2(y-4) = y^3 - 8 = 0 \Rightarrow y^3 = 8 \Rightarrow y = 2$.

$F'' = 3y^2 \Rightarrow F''(2) = 12 > 0 \Rightarrow$ The stationary point is minimum.

Hence, $(2, 2)$ is the point lies on the parabola $y^2 = 2x$ that is closest to the point $(1, 4)$. ∎

Example 9.21

Find the dimensions of the rectangle of largest area that can be inscribed in an equilateral triangle of side L if one side of the rectangle lie on the base of the triangle (Fig. 9.4).

Solution

Let the area A of the inscribed rectangle be $2xy$.

The height of the equilateral triangle is

$L^2 = \left(\dfrac{L}{2}\right)^2 + h^2 \Rightarrow h = \dfrac{\sqrt{3}}{2}L$.

By similar triangles, $\dfrac{y}{h} = \dfrac{\dfrac{L}{2} - x}{\dfrac{L}{2}} \Rightarrow y = \dfrac{h\left(\dfrac{L}{2} - x\right)}{\dfrac{L}{2}} = \sqrt{3}\left(\dfrac{L}{2} - x\right)$.

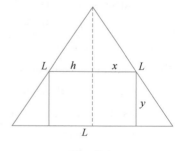

Fig. 9.4

$A = 2x\sqrt{3}\left(\dfrac{L}{2} - x\right) = 2\sqrt{3}\left(\dfrac{Lx}{2} - x^2\right) \Rightarrow$

$A' = 2\sqrt{3}\left(\dfrac{L}{2} - 2x\right) = 0 \Rightarrow x = \dfrac{L}{4}$.

$A'' = -4\sqrt{3} < 0 \Rightarrow$ The stationary point is maximum.

The rectangle has width $2x = \dfrac{L}{2}$, and the length of the rectangle is $y = \sqrt{3}\left(\dfrac{L}{2} - \dfrac{L}{4}\right) = \dfrac{\sqrt{3}L}{4}$. ∎

Summary of Key Theories 核心定义总结

(1) A function $f(x)$ is said to be increasing on an interval I, if $f(x_2) > f(x_1)$ whenever $x_2 > x_1$.

A function $f(x)$ is said to be decreasing on an interval I, if $f(x_2) < f(x_1)$ whenever $x_2 > x_1$.

A function $f(x)$ is said to be nonincreasing on an interval I, if $f(x_2) \leq f(x_1)$ whenever $x_2 > x_1$.

A function $f(x)$ is said to be nondecreasing on an interval I, if $f(x_2) \geq f(x_1)$ whenever $x_2 > x_1$.

(2) Suppose a function $f(x)$ is continuous on an interval I and differentiable on all the points in I except the endpoints.

If $f'(x) > 0$ in I, then $f(x)$ is increasing on I.

If $f'(x) < 0$ in I, then $f(x)$ is decreasing on I.

If $f'(x) \leq 0$ in I, then $f(x)$ is nonincreasing on I.

If $f'(x) \geq 0$ in I, then $f(x)$ is nondecreasing on I.

(3) A function $f(x)$ is said to have an absolute maximum value (or global maximum) $f(x_0)$ at the point x_0 in the domain of $f(x)$ if $f(x) \leq f(x_0)$ for all x in the domain of $f(x)$.

A function $f(x)$ is said to have an absolute minimum value (or global minimum) $f(x_0)$ at the point x_0 in the domain of $f(x)$ if $f(x) \geq f(x_0)$ for all x in the domain of $f(x)$.

A function $f(x)$ is said to have a local maximum value (or relative maximum) $f(x_0)$ at the point x_0 in the domain of $f(x)$ if $\exists h > 0$ such that $f(x) \leq f(x_0)$ whenever x is in the domain of $f(x)$ and $|x - x_0| < h$.

A function $f(x)$ is said to have a local minimum value (or relative minimum) $f(x_0)$ at the point x_0 in the domain of $f(x)$ if $\exists h > 0$ such that $f(x) \geq f(x_0)$ whenever x is in the domain of $f(x)$ and $|x - x_0| < h$.

(4) A stationary point of a function $f(x)$ is a point $(a, f(a))$ such that $f'(a) = 0$. The point $(a, f(a))$ could be a local maximum, local minimum or neither.

(5) Suppose $(a, f(a))$ is a stationary point for the function $f(x)$.

(a) If f' changes form positive to negative at $x = a$, then $(a, f(a))$ is a local maximum.

(b) If f' changes form negative to positive at $x = a$, then $(a, f(a))$ is a local minimum.

(c) If f' does not change signs at $x = a$ (positive to positive or negative to negative), then $(a, f(a))$ is neither a local maximum nor a local minimum.

(6) The function $f(x)$ is said to be concave up on an interval I, if $f'(x)$ is an increasing function on I. Similarly, a function $f(x)$ is concave down on an interval I, if $f'(x)$ is a decreasing function on I.

In other words, if the graph of $f(x)$ lies above all of its tangents on I, then it is concave upwards on I.

If the graph of $f(x)$ lies below all of its tangents on I, then it is concave downwards on I.

(7) Suppose $(a, f(a))$ is a stationary point for the function $f(x)$.

(a) If $f''(a) < 0$, then $(a, f(a))$ is a local maximum.

(b) If $f''(a) > 0$, then $(a, f(a))$ is a local minimum.

(c) If $f'' = 0$, then no conclusion; it may be an inflection point, but not guaranteed.

(8) A point P on a curve $y = f(x)$ is an inflection point, if $f(x)$ is continuous and the curve changes concavity at P (either concave upwards to downwards or vice versa).

Anti-derivative and Integration

第 10 章 反导数和积分

The second fundamental problem addressed by calculus is determining the area of a region bounded by curves. In order to solve such a problem, we need to use integration. The question is: What is integration, and how does it relate to the area bounded by curves? We begin this chapter with simple definition.

Definition 定义

An antiderivative of a function $f(x)$ on an interval I is another function $F(x)$ satisfying $F'(x) = f(x)$ $\forall x \in I$.

函数 $f(x)$ 在区间 I 的反导数是一个新函数 $F(x)$,满足 $F'(x) = f(x)$ $\forall x \in I$.

Example 例 10.1

(a) $F(x) = x$ is an antiderivative of $f(x) = 1$ on any interval because $F'(x) = 1$.
(b) $F(x) = x - 100$ is an antiderivative of $f(x) = 1$ on any interval because $F'(x) = 1$.
(c) $F(x) = x + 25$ is an antiderivative of $f(x) = 1$ on any interval because $F'(x) = 1$.

Solution 解

As we seen in Example 10.1, if $F(x)$ is an antiderivative of $f(x)$ so does $F(x) + C$, where C is a constant. Hence, $F(x) + C$ is the family of all antiderivatives of $f(x)$; and we called it the general antiderivative of $f(x)$.

Example 例 10.2

Find the general antiderivative of $f(x) = x^n$ for $n \neq -1$.

Solution 解

From the Power Rule, we know $\dfrac{d}{dx}\left(\dfrac{x^{n+1}}{n+1}\right) = \dfrac{n+1}{n+1}x^n = x^n$ if $n \neq -1$.

Chapter 10　Anti-derivative and Integration　第十章　反导数和积分

Therefore, $\dfrac{x^{n+1}}{n+1} + C$ is the general antiderivative of x^n if $n \neq -1$.

10.1　Indefinite Integral　不定积分

The general antiderivative of a function $f(x)$ on an interval I is $F(x) + C$, where $F(x)$ is any antiderivative of $f(x)$ and C is any constant.

Since $\dfrac{\mathrm{d}}{\mathrm{d}x}(F(x) + C) = f(x) \Rightarrow \mathrm{d}(F(x) + C) = f(x)\,\mathrm{d}x$.

We now introduce the symbol "$\int$" which represents the antidifferentiation. $F(x) + C = \int f(x)\,\mathrm{d}x$.

👑 Definition　定义

The indefinite integral of $f(x)$ on an interval I is denoted as $\int f(x)\,\mathrm{d}x = F(x) + C$, where $F'(x) = f(x)\ \forall x \in I$.

$f(x)$ 的不定积分在区间 I 表述为 $\int f(x)\,\mathrm{d}x = F(x) + C$，其中，$F'(x) = f(x)\ \forall x \in I$.

The symbol "$\int$" is called an integral sign, and we should treat $\int f(x)\,\mathrm{d}x$ as a single symbol representing the indefinite integral of $f(x)$ with respect to x. The constant C is called the constant of integration.

Since the anti-differentiation is the reverse process of differentiation, therefore, some basic rules can be obtained directly from differentiation.

📝 Basic Indefinite Integrals and Properties　基本不定积分和性质

$\int x^n\,\mathrm{d}x = \dfrac{1}{n+1}x^{n+1}$, for $n \neq -1$.

$\int c \cdot f(x)\,\mathrm{d}x = c \cdot \int f(x)\,\mathrm{d}x$.

$\int [f(x) \pm g(x)]\,\mathrm{d}x = \int f(x)\,\mathrm{d}x \pm \int g(x)\,\mathrm{d}x$.

✍ Example 例 10.3

Find x for the following equations.

(a) $\dfrac{\mathrm{d}x}{\mathrm{d}t} = t^2$;　(b) $\dfrac{\mathrm{d}x}{\mathrm{d}y} = y^3$;　(c) $\dfrac{\mathrm{d}x}{\mathrm{d}w} = w^{-5}$.

✋ Solution 解

(a) $x = \dfrac{1}{3}t^3 + C$;　(b) $x = \dfrac{1}{4}y^4 + C$;　(c) $x = \dfrac{-1}{4}w^{-4} + C$.

Example 例 10.4

Evaluate the followings:

(a) $\int (x^{\frac{2}{3}}) \, dx$; (b) $\int \left(\dfrac{1}{x^2} - \dfrac{1}{\sqrt{x}}\right) dx$; (c) $\int (x^2 - 3x + 5) \, dx$.

Solution 解

(a) $\int (x^{\frac{2}{3}}) \, dx = \dfrac{x^{\frac{2}{3}+1}}{\frac{2}{3}+1} + C = \dfrac{3}{5} x^{\frac{5}{3}} + C.$

(b) $\int \left(\dfrac{1}{x^2} - \dfrac{1}{\sqrt{x}}\right) dx = \int (x^{-2} - x^{\frac{-1}{2}}) \, dx = \dfrac{x^{-2+1}}{-2+1} - \dfrac{x^{\frac{-1}{2}+1}}{\frac{-1}{2}+1} + C = -x^{-1} - 2x^{\frac{1}{2}} + C.$

(c) $\int (x^2 - 3x + 5) \, dx = \dfrac{x^{2+1}}{2+1} - 3 \cdot \dfrac{x^{1+1}}{2} + 5x + C = \dfrac{1}{3} x^3 - \dfrac{3}{2} x^2 + 5x + C.$ ■

Example 例 10.5

Find the following indefinite integrals.

(a) $\int \dfrac{x^2 + 3x - 1}{\sqrt{x}} \, dx$; (b) $\int x(x+1)(x-2) \, dx$.

Solution 解

(a) $\int \dfrac{x^2 + 3x - 1}{\sqrt{x}} \, dx = \int \left(x^{\frac{3}{2}} + 3x^{\frac{1}{2}} - x^{\frac{-1}{2}}\right) dx = \dfrac{2}{5} x^{\frac{5}{2}} + 3 \cdot \dfrac{2}{3} x^{\frac{3}{2}} - 2x^{\frac{1}{2}} + C = \dfrac{2}{5} x^{\frac{5}{2}} + 2x^{\frac{3}{2}} - 2x^{\frac{1}{2}} + C.$

(b) $\int x(x+1)(x-2) \, dx = \int \left(x^3 - x^2 - 2x\right) dx = \dfrac{x^4}{4} - \dfrac{x^3}{3} - x^2 + C.$ ■

If we are given $\int (ax+b)^n \, dx$, where $a \neq 0$ and $n \neq -1$; we can use the idea of Chain Rule to obtain the expression, or we can use a substitution method to obtain the indefinite integral.

Let $u = ax + b \Rightarrow \dfrac{du}{dx} = a \Rightarrow du = a\, dx \Rightarrow dx = \dfrac{1}{a} du.$

$\int (ax+b)^n \, dx = \int u^n \cdot \dfrac{1}{a} \, du = \dfrac{1}{a} \int u^n \, du = \dfrac{1}{a} \cdot \dfrac{u^{n+1}}{n+1} + C = \dfrac{1}{a} \dfrac{(ax+b)^{n+1}}{n+1} + C.$

$$\boxed{\int (ax+b)^n \, dx = \dfrac{1}{a} \dfrac{(ax+b)^{n+1}}{n+1} + C \text{ where } a \neq 0 \text{ and } n \neq -1.}$$

Example 例 10.6

Find the following indefinite integrals.

(a) $\int (3x-5)^4 \, dx$; (b) $\int \dfrac{7}{(2-4x)^9} \, dx$; (c) $\int \dfrac{6}{\sqrt{5x-1}} \, dx$.

Chapter 10 Anti-derivative and Integration   第十章 反导数和积分

Solution 解

(a) Let $u = 3x - 5 \Rightarrow du = 3dx \Rightarrow \dfrac{1}{3}du = dx$.

$$\int (3x-5)^4 dx = \int u^4 \cdot \dfrac{1}{3} du = \dfrac{1}{3} \cdot \dfrac{1}{5} u^5 + C = \dfrac{1}{15}(3x-5)^5 + C.$$

(b) Let $u = 2 - 4x \Rightarrow du = -4dx \Rightarrow \dfrac{1}{-4}du = dx$.

$$\int \dfrac{7}{(2-4x)^9} dx = \int \dfrac{7}{u^9} \cdot \dfrac{-1}{4} du = \dfrac{-7}{4} \int u^{-9} du = \dfrac{-7}{4} \cdot \dfrac{1}{-8} u^{-8} + C = \dfrac{7}{32}(2-4x)^{-8} + C.$$

(c) Let $u = 5x - 1 \Rightarrow du = 5dx \Rightarrow \dfrac{1}{5}du = dx$.

$$\int \dfrac{6}{\sqrt{5x-1}} dx = \int \dfrac{6}{\sqrt{u}} \cdot \dfrac{1}{5} du = \dfrac{6}{5} \cdot 2u^{\frac{1}{2}} + C = \dfrac{12}{5}\sqrt{5x-1} + C.$$

Or you can use the formula directly.

(a) $\int (3x-5)^4 dx = \dfrac{1}{3} \dfrac{(3x-5)^5}{5} + C = \dfrac{(3x-5)^5}{15} + C.$

(b) $\int \dfrac{7}{(2-4x)^9} dx = 7\int (2-4x)^{-9} dx = 7 \cdot \dfrac{1}{-4} \cdot \dfrac{(2-4x)^{-8}}{-8} + C = \dfrac{7}{32}(2-4x)^{-8} + C.$

(c) $\int \dfrac{6}{\sqrt{5x-1}} dx = 6\int (5x-1)^{\frac{-1}{2}} dx = 6 \cdot \dfrac{1}{5} \dfrac{(5x-1)^{\frac{1}{2}}}{\frac{1}{2}} + C = \dfrac{12}{5}\sqrt{5x-1} + C.$ ∎

Example 例 10.7

Find the following indefinite integrals.

(a) $\int x(x+1)^{100} dx$;　　(b) $\int (x+1)^3 (x-4) dx$;　　(c) $\int (2x-3)^5 (4x+10) dx.$

Solution 解

(a) **Method 1**

If we try to use the binomial expansion to expand the term $(x+1)^{100}$, the calculation would be messy. So we simplify the expression first.

$x(x+1)^{100} = (x+1-1)(x+1)^{100} = (x+1)^{101} - (x+1)^{100}.$

$$\int x(x+1)^{100} dx = \int \left[(x+1)^{101} - (x+1)^{100} \right] dx = \dfrac{(x+1)^{102}}{102} - \dfrac{(x+1)^{101}}{101} + C.$$

Method 2

Let $u = x + 1 \Rightarrow du = dx$.

$$\int x(x+1)^{100} dx = \int (u-1)u^{100} du = \int (u^{101} - u^{100}) du = \dfrac{u^{102}}{102} - \dfrac{u^{101}}{101} + C = \dfrac{(x+1)^{102}}{102} - \dfrac{(x+1)^{101}}{101} + C.$$

(b) $\int (x+1)^3 (x-4) dx = \int (x+1)^3 (x+1-5) dx = \int \left[(x+1)^4 - 5(x+1)^3 \right] dx = \dfrac{(x+1)^5}{5} - 5\dfrac{(x+1)^4}{4} + C.$

(c) $\int (2x-3)^5(4x+10)\,dx = \int \left[(2x-3)^5(2(2x-3)+16)\right]dx = \int \left[2(2x-3)^6 + 16(2x-3)^5\right]dx$

$= 2\dfrac{(2x-3)^7}{2 \cdot 7} + 16\dfrac{(2x-3)^6}{2 \cdot 6} + C = \dfrac{(2x-3)^7}{7} + \dfrac{4}{3}(2x-3)^6 + C.$ ∎

Note 注意

(b) and (c) can be solved by using the method of substitution, we leave them for the readers to try.

It is common in the application of calculus that we are given $f'(x)$ ask to find the function $f(x)$.

An equation involves the derivatives of a function is called a differential equation which will be discuss in future course. At this stage, we can solve some simple differential equation.

The general solution of differential equations involves an arbitrary constant C as well. When extra conditions are given, we can determine the constants.

(b)和(c)可通过代入法求解，读者可以试一试.

在微积分的应用中，常需要通过 $f'(x)$ 来推导函数 $f(x)$. 涉及函数导数的方程可以称为微分方程，在未来的课程中我们会继续学习微分方程. 本章我们只需要掌握如何解基础的微分方程. 微分方程的通解包括一个任意常数 C. 如果有额外的给定条件，可以求出该常数值.

Example 例 10.8

The function $f(x)$ has derivative $f'(x) = x^2 + 3x - 5$ and it is given that $f(1) = 3$, find the expression of $f(x)$.

Solution 解

$f(x) = \int(x^2 + 3x - 5)\,dx = \dfrac{x^3}{3} + \dfrac{3}{2}x^2 - 5x + C.$

$f(1) = 3 \Rightarrow \dfrac{1}{3} + \dfrac{3}{2} - 5 + C = 3 \Rightarrow C = 6\dfrac{1}{6}.$

So $f(x) = \dfrac{x^3}{3} + \dfrac{3}{2}x^2 - 5x + 6\dfrac{1}{6}.$ ∎

Example 例 10.9

It is given that $\dfrac{dy}{dt} = 4t^3 + \sqrt{t} - 1$ and when $t = 1$, $y = 2$. Find the expression of y in terms of t.

Solution 解

$y = \int(4t^3 + \sqrt{t} - 1)\,dt = t^4 + \dfrac{2}{3}t^{\frac{3}{2}} - t + C.$

When $t = 1$, $y = 2 \Rightarrow 1 + \dfrac{2}{3} - 1 + C = 2 \Rightarrow C = \dfrac{4}{3}.$

Hence, $y = t^4 + \dfrac{2}{3}t^{\frac{3}{2}} - t + \dfrac{4}{3}.$ ∎

Example 例 10.10

A curve $y = f(x)$ has $\dfrac{dy}{dx} = 3x^2 - 2x + k$, where k is a constant. The equation of the tangent line to the curve at

Chapter 10　Anti-derivative and Integration

the point $(1,5)$ is $2x + y = 5$. Find the equation of the curve.

Solution 解

$2x + y = 5 \Rightarrow y = -2x + 5 \Rightarrow f'(1) = -2 \Rightarrow 3 - 2 + k = -2 \Rightarrow k = -3$.

$y = \int (3x^2 - 2x - 3) dx = x^3 - x^2 - 3x + C$, since $(1, 5)$ is on the curve. $1 - 1 - 3 + C = 5 \Rightarrow C = 8$.

So $y = x^3 - x^2 - 3x + 8$.

Example 例 10.11

A curve $y = f(x)$ has $\dfrac{d^2y}{dx^2} = 6x - 6$. The local maximum of the curve is $(-1, 7)$. Find the equation of the curve.

Solution 解

$\dfrac{d^2y}{dx^2} = 6x - 6 \Rightarrow \dfrac{dy}{dx} = \int (6x - 6) dx = 3x^2 - 6x + C_1$.

Since $(-1, 7)$ is a maximum, so it is a stationary point. $f'(-1) = 0 \Rightarrow 3 + 6 + C_1 = 0 \Rightarrow C_1 = -9$.

So $\dfrac{dy}{dx} = 3x^2 - 6x - 9 \Rightarrow y = \int (3x^2 - 6x - 9) dx = x^3 - 3x^2 - 9x + C \Rightarrow 7 = (-1)^3 - 3(-1)^2 - 9(-1) + C \Rightarrow C = 2$.

Therefore, $f(x) = x^3 - 3x^2 - 9x + 2$.

Example 例 10.12

A curve $y = f(x)$ is such that $\dfrac{dy}{dx} = 2x - 1$, the point $P(-1, 1)$ is on the curve. The normal line at P meets the curve again at the point Q. Find the coordinate of Q and the stationary point for the curve and determine the nature of the stationary point.

Solution 解

$\dfrac{dy}{dx} = 2x - 1 \Rightarrow y = \int (2x - 1) dx = x^2 - x + C$.

$P(-1, 1)$ is on the curve. $(-1)^2 - (-1) + C = 1 \Rightarrow C = -1$.

The curve is $y = x^2 - x - 1$. $y'' = 2 > 0 \Rightarrow$ The stationary point $\left(\dfrac{1}{2}, \dfrac{-5}{4}\right)$ is a minimum.

$y'(-1) = -2 - 1 = -3 \Rightarrow$ The gradient of the normal line is $\dfrac{-1}{-3} = \dfrac{1}{3}$.

The equation of the normal line is $\dfrac{y-1}{x+1} = \dfrac{1}{3} \Rightarrow y = \dfrac{1}{3}x + \dfrac{4}{3}$.

Solve the simultaneous equation $y = x^2 - x - 1$ and $y = \dfrac{1}{3}x + \dfrac{4}{3}$. $3x^2 - 4x - 7 = 0 \Rightarrow (x+1)(3x-7) = 0$.

The x-coordinate of Q is $\dfrac{7}{3}$, the y-coordinate of Q is $\dfrac{19}{9}$.

Example 10.13

Find the derivative of $y = \dfrac{3}{(x^2-5x+1)^4}$, hence, find $\displaystyle\int \dfrac{36(2x-5)}{(x^2-5x+1)^5}dx$.

Solution

$y = \dfrac{3}{(x^2-5x+1)^4} = 3(x^2-5x+1)^{-4} \Rightarrow y' = 3\cdot(-4)(x^2-5x+1)^{-5}(2x-5) = \dfrac{-12(2x-5)}{(x^2-5x+1)^5}.$

$\displaystyle\int \dfrac{36(2x-5)}{(x^2-5x+1)^5}dx = -3\int \dfrac{-12(2x-5)}{(x^2-5x+1)^5}dx = -3\cdot \dfrac{3}{(x^2-5x+1)^4} + C = \dfrac{-9}{(x^2-5x+1)^4} + C.$ ∎

Example 10.14

Find the derivative of $y = \dfrac{3}{\sqrt{(x^2+2x+5)}}$, hence, find $\displaystyle\int \dfrac{(x+1)}{\sqrt{(x^2+2x+5)^3}}dx$.

Solution

$y = \dfrac{3}{\sqrt{(x^2+2x+5)}} = 3(x^2+2x+5)^{\frac{-1}{2}}.$

$y' = 3\cdot \dfrac{-1}{2}(x^2+2x+5)^{\frac{-3}{2}}\cdot(2x+2) = \dfrac{-3(x+1)}{\sqrt{(x^2+2x+5)^3}}$

$\Rightarrow \displaystyle\int \dfrac{(x+1)}{\sqrt{(x^2+2x+5)^3}}dx = \dfrac{-1}{3}\cdot \dfrac{3}{\sqrt{(x^2+2x+5)}} + C = \dfrac{-1}{\sqrt{(x^2+2x+5)}} + C.$ ∎

* 10.2　Riemann Sum　黎曼积分

Before get into the definite integral of a function $f(x)$, we should introduce the Riemann Sum; even though it is not in the exam syllabus and the majority of the A-level textbook did not discuss it. It is a good idea to introduce the concept, so the students know why the definite integral represents the net area of region bounded by curves.

Suppose we are given a function $f(x)$, and we want to know what is the area bounded by the curve, x-axis, $x = a$ and $x = b$ (Fig. 10.1).

We know partitioned the region into n strips, and we now approximate the ith strip S_i by a rectangle with width $\Delta x = \dfrac{b-a}{n}$ and height $f(x_i^*)$, where x_i^* is any point in the interval (x_{i-1}, x_i). Therefore, the area of the ith rectangle is just equal to $f(x_i^*)\cdot \Delta x$. The area under the curve is approximately equal to

$f(x_1^*)\cdot \Delta x + f(x_2^*)\cdot \Delta x + \cdots + f(x_n^*)\cdot \Delta x.$

Chapter 10 Anti-derivative and Integration

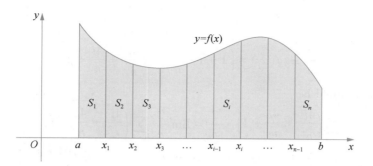

Fig. 10.1

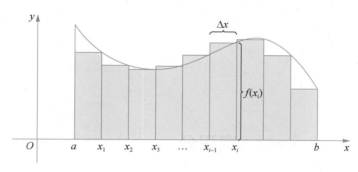

Fig. 10.2

If we have infinitely many rectangles, that is $n \to \infty$; then $\Delta x \to 0$ such that x_i^* will fall into a single point in the interval (x_{i-1}, x_i).

Definition 定义

The area A of a region S that lies under the graph of $f(x)$ is the limit of the sum of the area of approximating rectangles. $A = \lim_{n \to \infty}(f(x_1^*) \cdot \Delta x + f(x_2^*) \cdot \Delta x + \cdots + f(x_n^*) \cdot \Delta x)$.

我们把 $f(x)$ 图像下方 S 区域内的面积 A 称为近似矩形的面积之和的极限.
$A = \lim_{n \to \infty}(f(x_1^*) \cdot \Delta x + f(x_2^*) \cdot \Delta x + \cdots + f(x_n^*) \cdot \Delta x)$.

We often use sigma notation $\sum$ which means "sum them up", so

$$A = \lim_{n \to \infty}(f(x_1^*) \cdot \Delta x + f(x_2^*) \cdot \Delta x + \cdots + f(x_n^*) \cdot \Delta x) = \lim_{n \to \infty} \sum_{i=1}^{n} f(x_i^*) \cdot \Delta x.$$

There are some simple formulae that we should know.

$\sum_{i=1}^{n} 1 = n$, $\sum_{i=1}^{n} i = \dfrac{n(1+n)}{2}$, $\sum_{i=1}^{n} i^2 = \dfrac{n(n+1)(2n+1)}{6}$ and $\sum_{i=1}^{n} i^3 = \left[\dfrac{n(n+1)}{2}\right]^2$.

These formulae can be proved by the method of difference or by mathematical induction which will be covered in A-level further mathematic course.

Example 例 10.15

Find the area of the region bounded by $y = x^2$, the x-axis, $x = 1$ and $x = 3$.

Solution 解

$\Delta x = \dfrac{3-1}{n} = \dfrac{2}{n}$, $x_i = 1 + i \cdot \dfrac{2}{n} \Rightarrow f(x_i) = \left(1 + i \cdot \dfrac{2}{n}\right)^2 = \dfrac{4i^2}{n^2} + \dfrac{4i}{n} + 1.$

The area under the curve would be

$$\lim_{n \to \infty} \sum_{i=1}^{n} f(x_i^*) \cdot \Delta x = \lim_{n \to \infty} \sum_{i=1}^{n} \left(\dfrac{4i^2}{n^2} + \dfrac{4i}{n} + 1\right) \cdot \dfrac{2}{n} = \lim_{n \to \infty} \dfrac{2}{n} \sum_{i=1}^{n} \left(\dfrac{4i^2}{n^2} + \dfrac{4i}{n} + 1\right)$$

$$= \lim_{n \to \infty} \dfrac{2}{n} \left(\sum_{i=1}^{n} \dfrac{4i^2}{n^2} + \sum_{i=1}^{n} \dfrac{4i}{n} + \sum_{i=1}^{n} 1\right) = \lim_{n \to \infty} \dfrac{2}{n} \left(\dfrac{4}{n^2} \sum_{i=1}^{n} i^2 + \dfrac{4}{n} \sum_{i=1}^{n} i + \sum_{i=1}^{n} 1\right)$$

$$= \lim_{n \to \infty} \dfrac{2}{n} \left(\dfrac{4}{n^2} \cdot \dfrac{n(n+1)(2n+1)}{6} + \dfrac{4}{n} \cdot \dfrac{n(n+1)}{2} + n\right) = \dfrac{2 \cdot 4 \cdot 2}{6} + \dfrac{2 \cdot 4}{2} + 2$$

$$= \dfrac{16}{6} + 6 = 8\dfrac{2}{3}.$$

10.3 Definite Integral 定积分

From previous section, we know the area under the curve is equal to and
$$\lim_{n \to \infty}(f(x_1^*) \cdot \Delta x + f(x_2^*) \cdot \Delta x + \cdots + f(x_n^*) \cdot \Delta x).$$
We now give a name and notation for this limit.

Definition 定义

The definite integral is denoted by $\int_a^b f(x)\,dx$ which is equal to $\lim\limits_{n \to \infty}(f(x_1^*) \cdot \Delta x + f(x_2^*) \cdot \Delta x + \cdots + f(x_n^*) \cdot \Delta x)$ where $\Delta x = \dfrac{b-a}{n}$ which means the definite integral $\int_a^b f(x)\,dx$ is the "net" area of the region bounded by $f(x)$, x-axis, $x = a$ and $x = b$.

定积分的表达式为 $\int_a^b f(x)\,dx$，代表 $\lim\limits_{n \to \infty}(f(x_1^*) \cdot \Delta x + f(x_2^*) \cdot \Delta x + \cdots + f(x_n^*) \cdot \Delta x)$，其中，$\Delta x = \dfrac{b-a}{n}$，表示定积分 $\int_a^b f(x)\,dx$ 是 $f(x)$、x 轴、$x = a$ 和 $x = b$ 限定区域内的面积.

In the definite, we called the definite integral $\int_a^b f(x)\,dx$ represents the "net" area of the region bounded by $f(x)$, x-axis, $x = a$ and $x = b$ which means when the region is above x-axis it is a positive area while it is below the x-axis it is a negative area.

Example 例 10.16

Evaluate the following definite integrals:

(a) $\int_0^1 \sqrt{1-x^2}\,dx$;   (b) $\int_0^3 (x-1)\,dx$.

Solution 解

(a) We know the definite integral $\int_0^1 \sqrt{1-x^2}\,dx$ represent the area a quarter circle centered at the origin with radius 1 (Fig. 10.3). Hence, $\int_0^1 \sqrt{1-x^2}\,dx = \dfrac{\pi}{4}$.

(b) We know the definite integral $\int_0^3 (x-1)\,dx$ represent the "net" areas of the region bounded by the function, x-axis, $x = 0$ and as shown in Fig. 10.4. A_2 has area $\dfrac{1}{2}$ while A_1 has area 2, therefore $\int_0^3 (x-1)\,dx = 2 - \dfrac{1}{2} = \dfrac{3}{2}$.

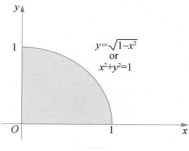
Fig. 10.3

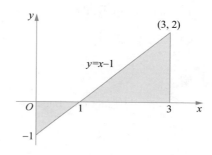
Fig. 10.4

Properties of Definite Integral　定积分的性质

Let $f(x)$ and $g(x)$ be two integrable functions on an interval I containing the points $x = a$, $x = b$ and $x = c$.

(a) $\int_a^a f(x)\,dx = 0$.

(b) $\int_a^b f(x)\,dx + \int_b^c f(x)\,dx = \int_a^c f(x)\,dx$.

(c) $\int_a^b f(x)\,dx = -\int_b^a f(x)\,dx$.

(d) $\int_a^b [k_1 f(x) \pm k_2 g(x)]\,dx = k_1 \int_a^b f(x)\,dx \pm k_2 \int_a^b g(x)\,dx$, where k_1 and k_2 are constants.

(e) $\int_{-a}^a f(x)\,dx = 0$ if $f(x)$ is an odd function.

(f) $\int_{-a}^a f(x)\,dx = 2\int_0^a f(x)\,dx$ if $f(x)$ is an even function.

假设 $f(x)$ 和 $g(x)$ 为包含 $x=a$，$x=b$ 和 $x=c$ 的区间 I 上的两个可积分函数.

(a) $\int_a^a f(x)\,dx = 0$.

(b) $\int_a^b f(x)\,dx + \int_b^c f(x)\,dx = \int_a^c f(x)\,dx.$

(c) $\int_a^b f(x)\,dx = -\int_b^a f(x)\,dx.$

(d) $\int_a^b [k_1 f(x) \pm k_2 g(x)]\,dx = k_1 \int_a^b f(x)\,dx \pm k_2 \int_a^b g(x)\,dx$,其中,$k_1$ 和 k_2 为常数.

(e) 如果 $f(x)$ 是奇函数,$\int_{-a}^a f(x)\,dx = 0.$

(f) 如果 $f(x)$ 是偶函数,$\int_{-a}^a f(x)\,dx = 2\int_0^a f(x)\,dx.$

Example 例 10.17

Evaluate the following definite integrals.

(a) $\int_{-3}^3 x^3\,dx$; (b) $\int_{-1}^1 \sqrt{1-x^2}\,dx.$

Solution 解

(a) Since $f(x) = x^3$ is an odd function, because $f(-x) = (-x)^3 = -x^3 = -f(x)$.

Therefore, $\int_{-3}^3 x^3\,dx = 0.$

(b) Since $f(x) = \sqrt{1-x^2}$ is an even function, because $f(-x) = \sqrt{1-(-x)^2} = \sqrt{1-x^2} = f(x)$.

Therefore, $\int_{-1}^1 \sqrt{1-x^2}\,dx = 2\int_0^1 \sqrt{1-x^2}\,dx = 2 \cdot \dfrac{\pi}{4} = \dfrac{\pi}{2}.$ ∎

At this point, we only know how to find definite integral by the means of "net" area. What if the regions are not the shapes which we know how to find their area?

We now introduce the Fundamental Theorem of Calculus which enable us to evaluate the definite integral without using the area of the region.

Theorem 10.1 Fundamental Theorem of Calculus (Part 1)
定理 10.1 微积分基本定理(第一部分)

Suppose $f(x)$ is continuous on an interval I containing the point $x = a$, for a function $F(x)$ defined as $F(x) = \int_a^x f(t)\,dt.$

The function $F(x)$ is differentiable on the interval I, and $\dfrac{d}{dx}\int_a^x f(t)\,dt = f(x).$

假设 $f(x)$ 在包含 $x = a$ 的区间 I 上是连续的,函数 $F(x)$ 定义如下:$F(x) = \int_a^x f(t)\,dt.$

那么,函数 $F(x)$ 在区间 I 上是可微的,并且 $\dfrac{d}{dx}\int_a^x f(t)\,dt = f(x).$

Proof 证明

We omit the proof here, because it required the knowledge of integral version of mean value theorem.

Example 例 10.18

Let $f(x) = \int_2^x (3t - 5) \, dt$, find the value of $f'(2)$.

Solution 解

$f'(x) = 3x - 5 \Rightarrow f'(2) = 6 - 5 = 1.$

Theorem 10.2 Fundamental Theorem of Calculus (Part 2)
定理 10.2 微积分基本定理(第二部分)

Suppose $f(x)$ is continuous on an interval I containing the points $x = a$ and $x = b$, then $\int_a^b f(x) \, dx = F(b) - F(a)$, where $F(x)$ is any antiderivative of $f(x)$.

假设 $f(x)$ 在包含 $x = a$ 和 $x = b$ 的区间 I 上是连续的,那么,$\int_a^b f(x) \, dx = F(b) - F(a)$,其中,$F(x)$ 是 $f(x)$ 的反导数.

Proof 证明

Let $g(x) = \int_a^x f(t) \, dt \Rightarrow g'(x) = f(x)$ which means $F(x) = g(x) + C$ where $F(x)$ is an antiderivative of $f(x)$ and C is a constant.

$g(a) = \int_a^a f(t) \, dt = 0$ and $g(b) = \int_a^b f(t) \, dt = F(b) - C$, also $F(a) = g(a) + C = C$.

$g(b) = \int_a^b f(t) \, dt = F(b) - C = F(b) - F(a)$.

We can now use the Fundamental of Calculus Part 2 together with Chain Rule to get the following results:

$\dfrac{d}{dx} \int_a^{g(x)} f(t) \, dt = \dfrac{d}{dx} [F(g(x)) - F(a)] = \dfrac{d}{dx} F(g(x)) = f(g(x)) \cdot g'(x)$ and

$\dfrac{d}{dx} \int_{h(x)}^{g(x)} f(t) \, dt = \dfrac{d}{dx} [F(g(x)) - F(h(x))] = \dfrac{d}{dx} F(g(x)) - \dfrac{d}{dx} F(h(x))$
$= f(g(x)) \cdot g'(x) - f(h(x)) \cdot h'(x).$

Example 例 10.19

Find the derivative of the following functions:

(a) $g(x) = \int_0^{x^2} \sqrt{1+t^2}\,dt$;   (b) $h(x) = \int_{x^2}^8 \sqrt{1+t^2}\,dt$;   (c) $k(x) = \int_{\sqrt{x}}^{x^2} \sqrt{1+t^2}\,dt$.

Solution 解

(a) $g'(x) = \sqrt{1+(x^2)^2} \cdot \dfrac{d}{dx}(x^2) = \sqrt{1+x^4} \cdot 2x$.

(b) $h(x) = \int_{x^2}^8 \sqrt{1+t^2}\,dt = -\int_8^{x^2} \sqrt{1+t^2}\,dt$.

$h'(x) = \sqrt{1+(x^2)^2} \cdot \dfrac{d}{dx}(x^2) = -\sqrt{1+x^4} \cdot 2x$.

(c) $k'(x) = \sqrt{1+(x^2)^2} \cdot \dfrac{d}{dx}(x^2) - \sqrt{1+(\sqrt{x})^2} \cdot \dfrac{d}{dx}(\sqrt{x}) = \sqrt{1+x^4} \cdot 2x - \dfrac{1}{2\sqrt{x}} \cdot \sqrt{1+x}$.

Example 例 10.20

If $\int_a^x f(t)\,dt = x^2 - x - 1$, find $f(x)$ and the value a.

Solution 解

$\dfrac{d}{dx}\left(\int_a^x f(t)\,dt\right) = \dfrac{d}{dx}(x^2 - x + 1) \Rightarrow f(x) = 2x - 1$.

$\int_a^a f(t)\,dt = 0 \Rightarrow a^2 - a - 1 = 0 \Rightarrow a = \dfrac{1 \pm \sqrt{1+4}}{2} = \dfrac{1 \pm \sqrt{5}}{2}$.

Example 例 10.21

Evaluate the following definite integrals：

(a) $\int_{-1}^3 (2x^2 - 3x + 1)\,dx$;   (b) $\int_1^4 \sqrt{3x+2}\,dx$.

Solution 解

(a) $\int_{-1}^3 (2x^2 - 3x + 1)\,dx = 2 \cdot \dfrac{x^3}{3} - 3 \cdot \dfrac{x^2}{2} + x \Big|_{-1}^3$

$= \dfrac{2}{3} \cdot 3^3 - \dfrac{3}{2} \cdot 3^2 + 3 - \left[\dfrac{2}{3} \cdot (-1)^3 - \dfrac{3}{2} \cdot (-1)^2 - 1\right] = \dfrac{32}{3}$.

(b) $\int_1^4 \sqrt{3x+2}\,dx = \dfrac{2}{3 \cdot 3}(3x+2)^{\frac{3}{2}} \Big|_1^4 = \dfrac{2}{9} \cdot 14^{\frac{3}{2}} - \dfrac{2}{9} \cdot 5^{\frac{3}{2}} = \dfrac{28}{9}\sqrt{14} - \dfrac{10}{9}\sqrt{5}$.

10.4 Area between Curves 曲线之间的面积

We know the area of the region bounded by the graph of $f(x)$, x-axis, $x = a$ and $x = b$ is defined as the definite integral $\int_a^b f(x)\,dx$. We now consider the region S lies between two continuous functions $y = f(x)$ and $y = g(x)$ and between two lines $x = a$ and $x = b$ such that $f(x) \geqslant g(x)$ in the interval $[a, b]$. We can use the Riemann Sum idea to find an approximation of the area of S.

The area of the rectangle as shown in the Fig. 10.5 is equal to $[f(x_i) - g(x_i)]\Delta x$, where $\Delta x = \dfrac{b-a}{n}$.

Hence, the area of $S = \lim\limits_{n\to\infty} \sum\limits_{i=1}^{n} [f(x_i) - g(x_i)]\Delta x = \int_a^b [f(x) - g(x)]\,dx$.

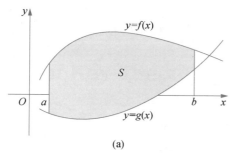
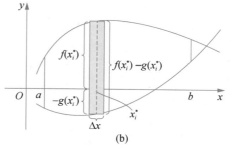

(a) (b)

Fig. 10.5

> The area of the region lies between two continuous functions $y = f(x)$ and $y = g(x)$ and between two lines $x = a$ and $x = b$, such that $f(x) \geqslant g(x)$ in the interval $[a, b]$ is equal to $\int_a^b (f(x) - g(x))\,dx$.
>
> 如果一个区域面积位于两个连续函数 $y = f(x)$ 和 $y = g(x)$ 之间，位于两条直线 $x = a$ 和 $x = b$ 之间，在区间 $[a, b]$，$f(x) \geqslant g(x)$ 等同于 $\int_a^b (f(x) - g(x))\,dx$。

Example 例 10.22

Find the area of the region bounded by $y = x^2 - 2x$ and $y = 4 - x^2$.

Solution 解

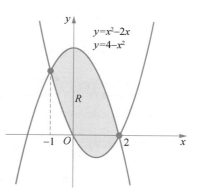

Fig. 10.6

We first sketch the graph as shown in Fig. 10.6. Therefore, we have to find the intersection points between two curves.

$x^2 - 2x = 4 - x^2 \Rightarrow 2x^2 - 2x - 4 = 0 \Rightarrow x^2 - x - 2 = 0 \Rightarrow (x-2)(x+1) = 0 \Rightarrow x = -1, 2$.

The area is equal to

$\int_{-1}^{2} [4 - x^2 - (x^2 - 2x)]\,dx = \int_{-1}^{2} (4 - 2x^2 + 2x)\,dx = 4x - \dfrac{2}{3}x^3 + x^2 \Big|_{-1}^{2}$

$= 4 \cdot 2 - \dfrac{2}{3} \cdot 2^3 + 2^2 - \left[4 \cdot -1 - \dfrac{2}{3}(-1)^3 + (-1)^2\right]$

$= 12 - \dfrac{16}{3} + \dfrac{7}{3} = 9$.

Example 例 <10.23>

Find the area bound by $y = \sqrt{x-2}$ and $x - y = 2$.

Solution 解

We first sketch the graph as shown in Fig. 10.7. Therefore, we have to find the intersection points between two curves.

$\sqrt{x-2} = x - 2 \Rightarrow x - 2 = (x-2)^2 \Rightarrow x^2 - 5x + 6 = 0 \Rightarrow (x-2)(x-3) = 0 \Rightarrow x = 2, 3$.

The area is equal to $\int_2^3 [\sqrt{x-2} - (x-2)]dx = \int_2^3 [\sqrt{x-2} - x + 2]dx = \frac{2}{3}(x-2)^{\frac{3}{2}} - \frac{x^2}{2} + 2x \Big|_2^3$

$= \frac{2}{3} - \frac{9}{2} + 6 - (0 - 2 + 4) = \frac{1}{6}$.

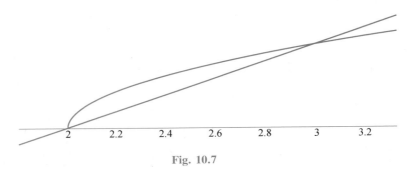

Fig. 10.7

Since we use vertical rectangles to find the area of the region bounded by curves, so the definite integral can be recognized as $\int_a^b y\,dx$. However, we can also use horizontal rectangles to find the area bounded by curves and the definite integral can be recognized as $\int_c^d x\,dy$. As shown in Fig. 10.8, the area bounded by two curves would be $\int_c^d [f(y) - g(y)]\,dy$.

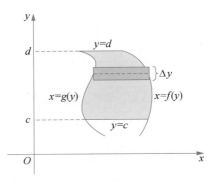

Fig. 10.8

Example 例 <10.24>

Find the area enclosed by $x = y^2 - 4y$ and $x = 2y - y^2$.

Solution 解

We first sketch the graph as shown in Fig. 10.9. Therefore, we have to find the intersection points between two curves.

$y^2 - 4y = 2y - y^2 \Rightarrow 2y^2 - 6y = 0 \Rightarrow 2y(y-3) = 0 \Rightarrow y = 0, 3$.

The area is equal to

$$\int_0^3 [2y - y^2 - (y^2 - 4y)]dy = \int_0^3 (6y - 2y^2)dy = 3y^2 - \frac{2}{3}y^3 \Big|_0^3 = 27 - 18 = 9.$$

Chapter 10 Anti-derivative and Integration

Each region A, B and C has area of 5 unit^2. Evaluate $\int_{-4}^{2}[2f(x)+3x-1]dx$.

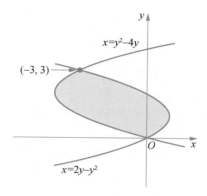

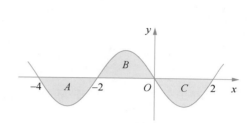

Fig. 10.9 Fig. 10.10

Solution

$\int_{-4}^{2}[2f(x)+3x-1)]dx = 2\int_{-4}^{2}f(x)dx + \int_{-4}^{2}(3x-1)dx = 2(-5+5-5) + \left(\dfrac{3}{2}x^2 - x\right)\Big|_{-4}^{2}$

$= -10 + \left\{\dfrac{3}{2}\cdot 2^2 - 2 - \left(\dfrac{3}{2}(-4)^2 - (-4)\right)\right\} = -10 + 6 - 2 - 28$

$= -4 - 30 = -34.$

10.5 Improper Integrals 广义积分

We have seen how to compute the definite integral in the form $\int_{a}^{b}f(x)dx$ where $f(x)$ is continuous and bounded on the closed interval $[a, b]$. Such definite integrals can also be called proper integrals. We are going to discuss the case that $a = -\infty$ or $b = +\infty$ or both, and the case that the function $f(x)$ is unbounded or discontinuous when x approaches a or b or both. Such integrals are called improper integrals.

Definition: Improper Integral of Type One 定义：第一类广义积分

If $f(x)$ is continuous on $[a, \infty)$, the integral $\int_{a}^{\infty}f(x)dx$ can be evaluated as $\lim\limits_{k\to\infty}\left(\int_{a}^{k}f(x)dx\right)$.

If $f(x)$ is continuous on $(-\infty, b]$, the integral $\int_{-\infty}^{b}f(x)dx$ can be evaluated as $\lim\limits_{k\to-\infty}\left(\int_{k}^{b}f(x)dx\right)$.

If $f(x)$ is continuous on $(-\infty, \infty)$, the integral $\int_{-\infty}^{\infty} f(x)\,dx$ can be evaluated as $\lim\limits_{k \to -\infty}\left(\int_{k}^{b} f(x)\,dx\right) + \lim\limits_{m \to \infty}\left(\int_{b}^{m} f(x)\,dx\right)$.

In the cases, if the limit exists and equal to a finite number, we said the improper integral is convergent; otherwise it is divergent. These types of integrals are called the improper integral of type 1.

如果 $f(x)$ 在区间 $[a, \infty)$ 是连续的,则积分 $\int_{a}^{\infty} f(x)\,dx$ 可被估算为 $\lim\limits_{k \to \infty}\left(\int_{a}^{k} f(x)\,dx\right)$.

如果 $f(x)$ 在区间 $(-\infty, b]$ 是连续的,则积分 $\int_{-\infty}^{b} f(x)\,dx$ 可被估算为 $\lim\limits_{k \to -\infty}\left(\int_{k}^{b} f(x)\,dx\right)$.

如果 $f(x)$ 在区间 $(-\infty, \infty)$ 是连续的,则积分 $\int_{-\infty}^{\infty} f(x)\,dx$ 可被估算为 $\lim\limits_{k \to -\infty}\left(\int_{k}^{b} f(x)\,dx\right) + \lim\limits_{m \to \infty}\left(\int_{b}^{m} f(x)\,dx\right)$.

在这些情况下,如果极限存在并等于一个有限数,则认为该广义积分是收敛的;否则就是发散的.这些类型的积分称作第一类广义积分.

Example 例 ⟨10.26⟩

Find the area of the region lying under $f(x) = \dfrac{1}{x^2}$ and above the x-axis to the right of $x = 1$.

Solution 解

The region is shown in Fig. 10.11.

So the area A of the region is $\int_{1}^{\infty} \dfrac{1}{x^2}\,dx$.

$\int_{1}^{\infty} \dfrac{1}{x^2}\,dx = \lim\limits_{k \to \infty}\left(\int_{1}^{k} \dfrac{1}{x^2}\,dx\right) = \lim\limits_{k \to \infty}\left(-x^{-1}\Big|_{1}^{k}\right) = \lim\limits_{k \to \infty}\left(\dfrac{-1}{k} + 1\right) = 1.$

So the improper integral is convergent, and the area of the unbounded region is equal to 1.

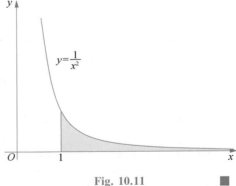

Fig. 10.11

Example 例 ⟨10.27⟩

It is given that $\dfrac{d}{dx}(\tan^{-1} x) = \dfrac{1}{1 + x^2}$. Find the area of the region lying under $f(x) = \dfrac{1}{1 + x^2}$ and above the x-axis.

Solution 解

The region is shown in Fig. 10.12. The area of the region is

$$\int_{-\infty}^{\infty} \dfrac{1}{1 + x^2}\,dx = \int_{-\infty}^{a} \dfrac{1}{1 + x^2}\,dx + \int_{a}^{\infty} \dfrac{1}{1 + x^2}\,dx.$$

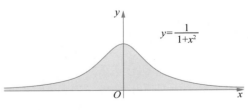

Fig. 10.12

We can choose a to be any finite number in the interval $(-\infty, +\infty)$, and we choose $a = 0$ for convenience.

$$\int_{-\infty}^{\infty} \frac{1}{1+x^2} dx = \int_{-\infty}^{0} \frac{1}{1+x^2} dx + \int_{0}^{\infty} \frac{1}{1+x^2} dx = \lim_{k \to -\infty} \left(\int_{k}^{0} \frac{1}{1+x^2} dx \right) + \lim_{m \to \infty} \left(\int_{0}^{m} \frac{1}{1+x^2} dx \right)$$

$$= \lim_{k \to -\infty} \left(\tan^{-1} x \Big|_{k}^{0} \right) + \lim_{m \to \infty} \left(\tan^{-1} x \Big|_{0}^{m} \right) = \lim_{k \to -\infty} (0 - \tan^{-1} k) + \lim_{m \to \infty} (\tan^{-1} m - 0) = \frac{\pi}{2} + \frac{\pi}{2} = \pi.$$

So the improper integral is convergent, and the area of the region is equal to π. ∎

Example 例 10.28

Determine the values of p such that the improper integral $\int_{1}^{\infty} \frac{1}{x^p} dx$ is convergent or divergent.

Solution 解

When $p = 1$, $\int_{1}^{\infty} \frac{1}{x} dx = \lim_{k \to \infty} \left(\ln x \Big|_{1}^{k} \right) = \lim_{k \to \infty} (\ln k) = \infty$, so it is divergent.

We will learn the function $f(x) = \ln x$ and the indefinite integral of $\frac{1}{x}$ in pure Mathematics 3. So we just take the result here.

When $p \neq 1$, $\int_{1}^{\infty} \frac{1}{x^p} dx = \lim_{k \to \infty} \left(\frac{x^{-p+1}}{-p+1} \Big|_{1}^{k} \right) = \lim_{k \to \infty} \left(\frac{k^{-p+1}}{-p+1} - \frac{1}{-p+1} \right)$.

If $p > 1$, $\lim_{k \to \infty} \left(\frac{k^{-p+1}}{-p+1} - \frac{1}{-p+1} \right) = 0 - \frac{1}{-p+1} = \frac{1}{p-1}$, so it is convergent.

If $p < 1$, $\lim_{k \to \infty} \left(\frac{k^{-p+1}}{-p+1} - \frac{1}{-p+1} \right) = \infty$, so it is divergent.

The improper integral $\int_{1}^{\infty} \frac{1}{x^p} dx$ is convergent when $p > 1$, and it is divergent when $p < 1$. ∎

Definition: Improper Integral of Type Two   定义:第二类广义积分

If $f(x)$ is continuous on $[a, b)$, possibly unbound near $x = b$ (discontinuous) then the integral $\int_{a}^{b} f(x) dx$ can be evaluated as $\lim_{k \to b^-} \left(\int_{a}^{k} f(x) dx \right)$ which means k is approaching b from left hand side.

If $f(x)$ is continuous on $(a, b]$, possibly unbound near $x = a$ (discontinuous) then the integral $\int_{a}^{b} f(x) dx$ can be evaluated as $\lim_{k \to a^+} \left(\int_{k}^{b} f(x) dx \right)$ which means k is approaching a from right hand side.

If $f(x)$ is continuous on (a, b), possibly unbound near $x = a$ (discontinuous) and $x = a$ (discontinuous), then the integral $\int_{a}^{b} f(x) dx$ can be evaluated as $\lim_{k \to a^+} \left(\int_{k}^{c} f(x) dx \right) + \lim_{m \to b^-} \left(\int_{c}^{m} f(x) dx \right)$ for any $c \in (a, b)$.

In the cases, if the limit exists and equal to a finite number, we said the improper integral is convergent; otherwise it is divergent. These types of integrals are called the improper integral of type 2.

如果 $f(x)$ 在区间 $[a, b)$ 是连续的，那么，该函数在近 $x = b$（不连续）可能无界，则积分 $\int_a^b f(x)\,dx$ 可被估算为 $\lim\limits_{k \to b^-}\left(\int_a^k f(x)\,dx\right)$，表示 k 从左侧接近 b.

如果 $f(x)$ 在区间 $(a, b]$ 是连续的，那么，该函数在近 $x = a$（不连续）可能无界，则积分 $\int_a^b f(x)\,dx$ 可被估算为 $\lim\limits_{k \to a^+}\left(\int_k^b f(x)\,dx\right)$，表示 k 从右侧接近 a.

如果 $f(x)$ 在区间 (a, b) 是连续的，那么，该函数在近 $x = b$（不连续）和近 $x = a$（不连续）可能无界，则积分 $\int_a^b f(x)\,dx$ 可被估算为 $\lim\limits_{k \to a^+}\left(\int_k^c f(x)\,dx\right) + \lim\limits_{m \to b^-}\left(\int_c^m f(x)\,dx\right)$，对所有 $c \in (a, b)$ 成立, 表示 k 从右侧接近 a.

在这些情况下，如果极限存在并等于一个有限数，则认为该广义积分是收敛的；否则就是发散的. 这些类型的积分称作第二类广义积分.

Example 例 10.29

Find the area of the unbounded region lying under $y = \dfrac{1}{\sqrt{x}}$ above the x-axis, between $x = 0$ and $x = 1$.

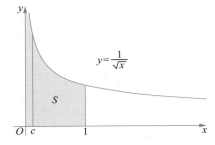

Fig. 10.13

Solution 解

The unbounded region is shown in Fig. 10.13.
So the area of the region S is equal to

$$\int_0^1 \frac{1}{\sqrt{x}}\,dx = \lim_{c \to 0^+}\left(\int_c^1 \frac{1}{\sqrt{x}}\,dx\right) = \lim_{c \to 0^+}\left(2x^{\frac{1}{2}}\,\Big|_c^1\right) = \lim_{c \to 0^+}(2 - 2c^{\frac{1}{2}}) = 2.$$

Example 例 10.30

Evaluate $\int_0^3 \dfrac{1}{\sqrt{3 - x}}\,dx$.

Solution 解

The function $f(x) = \dfrac{1}{\sqrt{3 - x}}$ is not defined at $x = 3$, so

$$\int_0^3 \frac{1}{\sqrt{3 - x}}\,dx = \lim_{k \to 3^-}\left(\int_0^k \frac{1}{\sqrt{3 - x}}\,dx\right) = \lim_{k \to 3^-}\left(-2(3 - x)^{\frac{1}{2}}\,\Big|_0^k\right) = \lim_{k \to 3^-}(-2(3 - k)^{\frac{1}{2}} - (-2(3 - 0)^{\frac{1}{2}}))$$
$$= 0 + 2\sqrt{3} = 2\sqrt{3}.$$

Chapter 10　Anti-derivative and Integration　　第十章　反导数和积分

10.6　Volume of Revolution　旋转体体积

We are going to find the volume of solid which is obtained by taking the region bounded by a curve, x-axis, $x = a$ and $x = b$ rotated by x-axis or y-axis, or the region bounded by a curve, y-axis, $y = c$ and $y = d$ rotated by x-axis or y-axis.

Definition　定义

Let S be s solid lying between $x = a$ and $x = b$, if the cross sectional area of S in the xy-plane is perpendicular to x-axis and it is a function $f(x)$, then the volume V of the solid is equal to

$$V = \lim_{n \to \infty} \left(\sum_{i=1}^{n} f(x_i^*) \cdot \Delta x \right) = \int_a^b f(x) \, dx.$$

If S is a solid lying between $y = c$ and $y = d$, if the cross sectional area of S in the xy-plane is perpendicular to y-axis and it is a function $g(y)$, then the volume V of the solid is equal to

$$V = \lim_{n \to \infty} \left(\sum_{i=1}^{n} g(y_i^*) \cdot \Delta y \right) = \int_c^d g(x) \, dy.$$

假设 S 是在 $x = a$ 和 $x = b$ 之间的一个立方体，如果 S 在 xy 平面上的横截面垂直于 x 轴，该面积函数为 $f(x)$，那么，该立方体的体积 $V = \lim\limits_{n \to \infty} \left(\sum\limits_{i=1}^{n} f(x_i^*) \cdot \Delta x \right) = \int_a^b f(x) \, dx.$

假设 S 是在 $y = c$ 和 $y = d$ 之间的一个立方体，如果 S 在 xy 平面上的横截面垂直于 y 轴，该面积函数为 $g(y)$，那么，该立方体的体积 $V = \lim\limits_{n \to \infty} \left(\sum\limits_{i=1}^{n} g(y_i^*) \cdot \Delta y \right) = \int_c^d g(x) \, dy.$

There are several methods for finding the volume of solid of revolution, however we are only focus on the "disk" and "washer" method. These method enable us to find the cross sectional area as a function $f(x)$ or $g(y)$.

Definition: Disk Method　定义：圆盘法

If the region A bound by $f(x)$, x-axis, $x = a$ and $x = b$ is rotated about the x-axis, then the cross sectional area of the solid generated is perpendicular to x-axis at any point between $x = a$ and $x = b$ is a circular disk. Therefore, the function of the cross sectional area is $\pi(f(x))^2$; hence, the volume of the solid is $\int_a^b \pi [f(x)]^2 \, dx$.

Similarly, if the region A bound by $g(y)$, y-axis, $y = c$ and $y = d$ is rotated about the y-axis, then the cross sectional area of the solid generated is perpendicular to y-axis at any point between $y = c$ and $y = d$ is a circular disk. Therefore, the function of the cross sectional area is $\pi [g(y)]^2$; hence, the volume of the solid is $\int_c^d \pi[g(y)]^2 dy$.

若区域 A 被 $f(x)$、x 轴、$x = a$ 和 $x = b$ 包围, 当区域 A 关于 x 轴旋转, 形成的立方体的横截面是一个圆盘, 并与 $x = a$ 和 $x = b$ 之间 x 轴上任意一点垂直. 该横截面的面积函数是 $\pi[f(x)]^2$, 该立方体的体积是 $\int_a^b \pi[f(x)]^2 dx$.

同理, 若区域 A 被 $g(y)$、y 轴、$y = c$ 和 $y = d$ 包围, 当区域 A 关于 y 轴旋转, 形成的立方体的横截面是一个圆盘, 并与 $y = c$ 和 $y = d$ 之间 y 轴上任意一点垂直. 该横截面的面积函数是 $\pi[g(y)]^2$, 该立方体的体积是 $\int_c^d \pi[g(y)]^2 dy$.

Example 例 10.31

Find the volume of the solid of revolution by taking the region bounded by the curve $y = \sqrt{x}$, x-axis, $x = 0$ and $x = 1$ rotated about the x-axis (Fig. 10.14).

Solution 解

The cross sectional area perpendicular to the axis of revolution is a circle with radius $y = \sqrt{x}$.

So the cross section area is a function $f(x) = \pi(\sqrt{x})^2 = \pi x$.

The volume is $V = \int_0^1 \pi x \, dx = \pi \dfrac{x^2}{2} \bigg|_0^1 = \dfrac{\pi}{2}$.

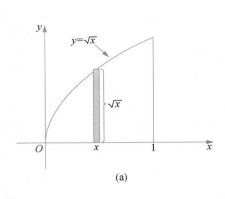
(a)

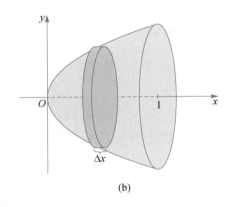
(b)

Fig. 10.14

Example 例 10.32

Find the volume of the solid of revolution by taking the region bounded by the curve $y = x^3$, y-axis, $y = 0$ and $y = 8$ rotated about the y-axis (Fig. 10.15).

Solution 解

The cross sectional area perpendicular to the axis of revolution is a circle with radius $x = y^{\frac{1}{3}}$.

Chapter 10 Anti-derivative and Integration

So the cross section area is a function $f(y) = \pi(y^{\frac{1}{3}})^2 = \pi y^{\frac{2}{3}}$.

The volume is $V = \int_0^8 \pi y^{\frac{2}{3}} dy = \pi \dfrac{3y^{\frac{5}{3}}}{5} \bigg|_0^8 = \dfrac{3\pi}{5} \cdot 2^5 = \dfrac{96\pi}{5}$.

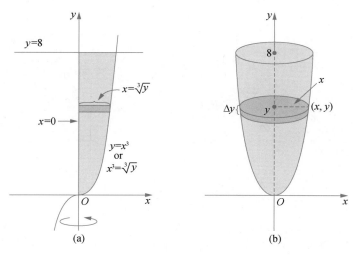

Fig. 10.15

Definition: Washer Method   定义:垫圈法

If the region A bound by $f(x)$, $g(x)$, $x = a$ and $x = b$ where $f(x) \geq g(x)$ on the interval $[a, b]$, is rotated about the x-axis, then the cross sectional area of the solid generated is perpendicular to x-axis at any point between $x = a$ and $x = b$, and it is an annulus. Therefore, the function of the cross sectional area is $\pi[f(x)]^2 - \pi[g(x)]^2$; hence, the volume of the solid is $\int_a^b \pi[f(x)]^2 - \pi[g(x)]^2 dx$.

Similarly, if the region A bound by $f(y)$, $g(y)$, $y = c$ and $y = d$ where $f(y) \geq g(y)$ on the interval $[c, d]$ is rotated about the y-axis, then the cross sectional area of the solid generated is perpendicular to y-axis at any point between $y = c$ and $y = d$, and it is an annulus. Therefore, the function of the cross sectional area is $\pi[f(y)]^2 - \pi[g(y)]^2$; hence, the volume of the solid is $\int_c^d \pi[f(y)]^2 - \pi[g(y)]^2 dy$.

若区域 A 被 $f(x)$, $g(x)$, $x=a$ 和 $x=b$ 包围,并且在区间 $[a,b]$ 内 $f(x) \geq g(x)$,当区域 A 关于 x 轴旋转,形成的立方体的横截面是一个圆环,并与 $x=a$ 和 $x=b$ 之间 x 轴上任意一点垂直.该横截面的面积函数是 $\pi[f(x)]^2 - \pi[g(x)]^2$,该立方体的体积是 $\int_a^b \{\pi[f(x)]^2 - \pi[g(x)]^2\} dx$.

同理,若区域 A 被 $f(y)$, $g(y)$, $y=c$ 和 $y=d$ 包围,并且在区间 $[c,d]$ 内 $f(y) \geq g(y)$,当区域 A 关于 y 轴旋转,形成的立方体的横截面是一个圆环,并与 $y=c$ 和 $y=d$ 之间 y 轴上的任意一点垂直.该横截面的面积函数是 $\pi[f(y)]^2 - \pi[g(y)]^2$,该立方体的体积是 $\int_c^d \{\pi[f(y)]^2 - \pi[g(y)]^2\} dy$.

Example 10.33

Find the volume of revolution by taking the region bounded by $y = x$ and $y = x^2$ rotated about x-axis (Fig. 10.16).

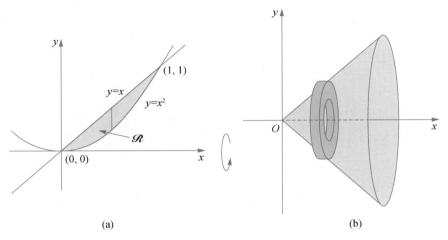

(a)　　　(b)

Fig. 10.16

Solution

We first solve for the intersection points between curves.
$x = x^2 \Rightarrow x^2 - x = 0 \Rightarrow x(x-1) = 0 \Rightarrow x = 0, 1.$

The volume $V = \int_0^1 [\pi(x)^2 - \pi(x^2)^2]\,dx = \pi\int_0^1 (x^2 - x^4)\,dx = \pi\left(\dfrac{x^3}{3} - \dfrac{x^5}{5}\right)\bigg|_0^1 = \dfrac{2}{15}\pi.$ ∎

Example 10.34

Find the volume of revolution by taking the region bounded by $y = x$ and $y = x^2$ rotated about the horizontal line $y = 3$.

Solution

Refer to the figure in example 10.32, we know the outer radius is $3 - x^2$, and the inner radius is $3 - x$.

The volume $V = \pi\int_0^1 [(3-x^2)^2 - (3-x)^2]\,dx = \pi\int_0^1 [x^4 - 6x^2 + 9 - (x^2 - 6x + 9)]\,dx$

$= \pi\int_0^1 [x^4 - 7x^2 + 6x]\,dx = \pi\left(\dfrac{x^5}{5} - \dfrac{7}{3}x^3 + 3x^2\right)\bigg|_0^1 = \dfrac{13}{15}\pi.$ ∎

Example 10.35

Find the volume of revolution by taking the region bounded by $y = x$ and $y = x^2$ rotated about the horizontal line $y = -1$.

Solution

Refer to the figure in example 10.32, we know the outer radius is $1 + x$, and the inner radius is $1 + x^2$.

$$V = \pi\int_0^1 [(1+x)^2 - (1+x^2)^2]dx = \pi\int_0^1 [x^2 + 2x + 1 - (x^4 + 2x^2 + 1)]dx$$
$$= \pi\int_0^1 (-x^4 - x^2 + 2x)dx = \pi\left(-\frac{x^5}{5} - \frac{1}{3}x^3 + x^2\right)\bigg|_0^1 = \frac{4}{15}\pi.$$

Example 10.36

Find the volume of revolution by taking the region bounded by $y = x$ and $y = x^2$ rotated about the vertical line $x = -1$.

Solution

Refer to the figure in example 10.32, we know the outer radius is $1 + x = 1 + \sqrt{y}$, and the inner radius is $1 + x = 1 + y$. When $x = 0 \Rightarrow y = 0$ and $x = 1 \Rightarrow y = 1$.

The volume $V = \pi\int_0^1 [(1+\sqrt{y})^2 - (1+y)^2]dy = \pi\int_0^1 (2\sqrt{y} - y - y^2)dy = \frac{\pi}{2}.$

Example 10.37

Find the volume of revolution by taking the region bounded by $y = x$ and $y = x^2$ rotated about the vertical line $x = 3$.

Solution

Refer to the figure in example 10.32, we know the outer radius is $3 - x = 3 - y$, and the inner radius is $3 - x = 3 - \sqrt{y}$. When $x = 0 \Rightarrow y = 0$ and $x = 1 \Rightarrow y = 1$.

The volume $V = \pi\int_0^1 [(3-y)^2 - (3-\sqrt{y})^2]dy = \pi\int_0^1 (y^2 - 7y + 6\sqrt{y})dy = \frac{5\pi}{6}.$

Summary of Key Theories　核心定义总结

(1) An antiderivative of a function $f(x)$ on an interval I is another function $F(x)$ satisfying $F'(x) = f(x)$ $\forall x \in I$.

(2) The indefinite integral of $f(x)$ on an interval I is denoted as $\int f(x)dx = F(x) + C$, where $F'(x) = f(x)$ $\forall x \in I$.

(3) $\int x^n dx = \frac{1}{n+1}x^{n+1}$, for $n \neq -1$.

$\int c \cdot f(x)dx = c \cdot \int f(x)dx.$

$\int [f(x) \pm g(x)]dx = \int f(x)dx \pm \int g(x)dx.$

(4) The area A of a region S that lies under the graph of $f(x)$ is the limit of the sum of the area of approximating rectangles. $A = \lim_{n \to \infty} [f(x_1^*) \cdot \Delta x + f(x_2^*) \cdot \Delta x + \cdots + f(x_n^*) \cdot \Delta x]$.

(5) The definite integral is denoted by $\int_a^b f(x) \, dx$ which is equal to $\lim_{n \to \infty} [f(x_1^*) \cdot \Delta x + f(x_2^*) \cdot \Delta x + \cdots + f(x_n^*) \cdot \Delta x]$ where $\Delta x = \dfrac{b - a}{n}$ which means the definite integral $\int_a^b f(x) \, dx$ is the "net" area of the region bounded by $f(x)$, x-axis, $x = a$ and $x = b$.

(6) Let $f(x)$ and $g(x)$ be two integrable functions on an interval I containing the points $x = a$, $x = b$ and $x = c$. Then

(a) $\int_a^a f(x) \, dx = 0$.

(b) $\int_a^b f(x) \, dx + \int_b^c f(x) \, dx = \int_a^c f(x) \, dx$.

(c) $\int_a^b f(x) \, dx = -\int_b^a f(x) \, dx$.

(d) $\int_a^b [k_1 f(x) \pm k_2 g(x)] \, dx = k_1 \int_a^b f(x) \, dx \pm k_2 \int_a^b g(x) \, dx$, where k_1 and k_2 are constants.

(e) $\int_{-a}^a f(x) \, dx = 0$ if $f(x)$ is an odd function.

(f) $\int_{-a}^a f(x) \, dx = 2 \int_0^a f(x) \, dx$ if $f(x)$ is an even function.

(7) Suppose $f(x)$ is continuous on an interval I containing the point $x = a$, for a function $F(x)$ defined as $F(x) = \int_a^x f(t) \, dt$.

The function $F(x)$ is differentiable on the interval I, and $\dfrac{d}{dx} \int_a^x f(t) \, dt = f(x)$.

Suppose $f(x)$ is continuous on an interval I containing the points $x = a$ and $x = b$, then $\int_a^b f(x) \, dx = F(b) - F(a)$, where $F(x)$ is any antiderivative of $f(x)$.

(8) If $f(x)$ is continuous on $[a, +\infty)$, the integral $\int_a^{+\infty} f(x) \, dx$ can be evaluated as $\lim_{k \to +\infty} \left(\int_a^k f(x) \, dx \right)$.

If $f(x)$ is continuous on $(-\infty, b]$, the integral $\int_{-\infty}^b f(x) \, dx$ can be evaluated as $\lim_{k \to -\infty} \left(\int_k^b f(x) \, dx \right)$.

If $f(x)$ is continuous on $(-\infty, +\infty)$, the integral $\int_{-\infty}^{\infty} f(x) \, dx$ can be evaluated as $\lim_{k \to -\infty} \left(\int_k^b f(x) \, dx \right) + \lim_{m \to +\infty} \left(\int_b^m f(x) \, dx \right)$.

In the cases, if the limit exists and equal to a finite number, we said the improper integral is convergent; otherwise it is divergent. These types of integrals are called the improper integral of type 1.

(9) If $f(x)$ is continuous on $[a, b)$, possibly unbound near $x = b$ (discontinuous) then the integral $\int_a^b f(x) \, dx$ can be evaluated as $\lim_{k \to b^-} \left[\int_a^k f(x) \, dx \right]$ which means k is approaching b from left hand side.

Chapter 10 Anti-derivative and Integration   第十章 反导数和积分

If $f(x)$ is continuous on $(a, b]$, possibly unbound near $x = a$ (discontinuous) then the integral $\int_a^b f(x)\,dx$ can be evaluated as $\lim_{k \to a^+} \left[\int_k^b f(x)\,dx\right]$ which means k is approaching a from right hand side.

If $f(x)$ is continuous on (a, b), possibly unbound near $x = b$ (discontinuous) and $x = a$ (discontinuous), then the integral $\int_a^b f(x)\,dx$ can be evaluated as $\lim_{k \to a^+}\left[\int_k^c f(x)\,dx\right] + \lim_{m \to b^-}\left[\int_c^m f(x)\,dx\right]$ for any $c \in (a, b)$.

In the cases, if the limit exists and equal to a finite number, we said the improper integral is convergent; otherwise it is divergent. These types of integrals are called the improper integral of type 2.

(10) Let S be s solid lying between $x = a$ and $x = b$, if the cross sectional area of S in the xy-plane is perpendicular to x-axis and it is a function $f(x)$, then the volume V of the solid is equal to
$$V = \lim_{n \to \infty}\left(\sum_{i=1}^n f(x_i^*) \cdot \Delta x\right) = \int_a^b f(x)\,dx.$$

If S is a solid lying between $y = c$ and $y = d$, if the cross sectional area of S in the xy-plane is perpendicular to y-axis and it is a function $g(y)$, then the volume V of the solid is equal to
$$V = \lim_{n \to \infty}\left(\sum_{i=1}^n g(y_i^*) \cdot \Delta y\right) = \int_c^d g(y)\,dy.$$

(11) If the region A bound by $f(x)$, x-axis, $x = a$ and $x = b$ is rotated about the x-axis, then the cross sectional area of the solid generated is perpendicular to x-axis at any point between $x = a$ and $x = b$ is a circular disk. Therefore, the function of the cross sectional area is $\pi[f(x)]^2$; hence, the volume of the solid is $\int_a^b \pi[f(x)]^2\,dx$.

Similarly, if the region A bound by $g(y)$, y-axis, $y = c$ and $y = d$ is rotated about the y-axis, then the cross sectional area of the solid generated is perpendicular to y-axis at any point between $y = c$ and $y = d$ is a circular disk. Therefore, the function of the cross sectional area is $\pi[g(y)]^2$; hence, the volume of the solid is $\int_c^d \pi[g(y)]^2\,dy$.

(12) If the region A bound by $f(x)$, $g(x)$, $x = a$ and $x = b$ where $f(x) \geq g(x)$ on the interval $[a, b]$, is rotated about the x-axis, then the cross sectional area of the solid generated is perpendicular to x-axis at any point between $x = a$ and $x = b$, and it is an annulus. Therefore, the function of the cross sectional area is $\pi[f(x)]^2 - \pi[g(x)]^2$; hence, the volume of the solid is $\int_a^b \{\pi[f(x)]^2 - \pi[g(x)]^2\}\,dx$.

Similarly, if the region A bound by $f(y)$, $g(y)$, $y = c$ and $y = d$ where $f(y) \geq g(y)$ on the interval $[c, d]$ is rotated about the y-axis, then the cross sectional area of the solid generated is perpendicular to y-axis at any point between $y = c$ and $y = d$, and it is an annulus. Therefore, the function of the cross sectional area is $\pi[f(y)]^2 - \pi[g(y)]^2$; hence, the volume of the solid is
$$\int_c^d \pi[f(y)]^2 - \pi[g(y)]^2\,dy.$$

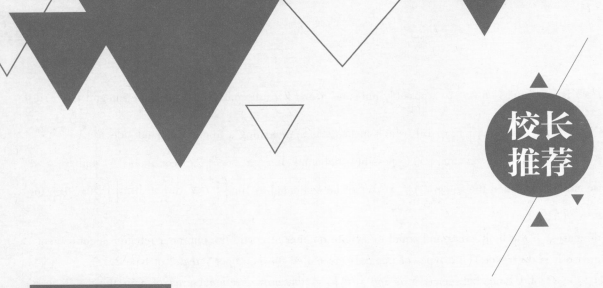

Stephen Tema 校长

The intention of this book is to prepare our students to attain a level which allows them to compete with other students applying for a G5 university as well as Oxbridge applications. The Mathematics curriculum is very specific and detailed and the content of this book sets the level which all students must reach and understand in order to satisfy the criteria for applying for a G5 university in this discipline of Mathematics. The author of this book has been working closely with admission offices of these G5 institutions and has prepared this book so senior students of the Ray Education Group may work towards reaching this level. The author is currently a HOD and senior member of the teaching staff at Adcote School in Shanghai.

Stephen Tema

Deputy Principal of Adcote School (Shanghai)

黄利鸣校长

It is my pleasure to introduce to you the Cambridge A-level International Course Preparation Book Series, compiled by Ray Education Group. The book series spans across multiple subjects and is aimed to be an informative, approachable, and helpful guide for students. Most importantly, it is a resource based closely on the standardized Cambridge A-level Syllabus, hence the series title.

What distinguishes our series is that it is a hybrid text, made specifically to help Chinese students learning in International A-level schools. Each book is written entirely in English, but comprehensive Chinese translations of key definitions and conclusions are provided by our professional team at Ray Education Group. All definitions are based strictly on the standardized Cambridge A-level Syllabus to ensure consistency and correctness.

One of the goals of the series is to simplify the Cambridge A-level Syllabus to the most essential knowledge points, allowing students of all levels to access the resource on the most fundamental level. Along with key terms and definitions, all the books that are in the series will contain examples and sample questions.

Students can expect these books to be a useful resource to be used in support of classroom textbooks, rather than in place of them. Altogether, there is indeed a rich concentration of knowledge of multiple subjects in A-level to be found across the series.

Liming Huang

Chief Academic Officer, Ray Education Group

Principal of A-level Center at Litai College, SISU

学生推荐

程照秋

Steve 的讲义陪伴我度过这个冬天约 80% 的漫长时光,它真的太方便我复习了!当生活疲劳乏味的时候,复习讲义就会瞬间快乐、充实起来!从最开始的基本原理到后面的题目,每一节知识都像一级台阶一样一点点铺垫起来,这让我在刷题的时候根本没有困难!而且每一章节复习完之后,知识存储在大脑里非常有成就感!总之,Steve 的讲义真的非常适合精心钻研!现在 Steve 编写的数学及进阶数学的书籍即将出版,真的是 A-level 学生的福音.

Sabrina(程照秋)

录取院校:牛津大学(工程)

石 立

作为 Steve 的学生,他的讲义使我受益匪浅.现在讲义即将出版成书,这本书深入浅出地解释了每一个可能会遇到的知识点,同时解释了其背后的数学原理,单从这一点就能感受到 Steve 对数学的理解.对于 A-level 学生来说,它是一本完全可以代替教材的辅导书.

Lili(石立)

录取院校:牛津大学(物理)

王忻楠

周老师出的书超级实用,全面针对最新的数学以及高数考纲.有的时候因为有事缺课,只要看看周老师的讲义就能保证全都搞懂,考试也能几乎全对.周老师在讲解如何解题的同时,还会详细地推导公式,让数学非常有趣.因为我有好多朋友考前喜欢背公式,但其实了解了怎么推导之后,考试的时候只需十几秒推导一下公式就能做题,全能的周老师会用最快、最棒的方式带你进入最快乐的数学世界!

Annie(王忻楠) CAIE A-level 数学(9709)满分

录取院校:剑桥大学(工程)

闫星宇

身为过来人才知道,一本好的教材有多么重要.全面的知识点,仔细的讲解,十分适合学生用来学习数学.如果你要申请牛剑,那你选对了——很有可能你的面试题就隐藏在其中;如果你不申请牛剑,那你也选对了——没有一套书能做到如此适合自学考试.周老师所编著的书是我到现在为止见过的最好的、最全的、最适合学习 A-level 数学以及进阶数学的辅导用书.

Adrian(闫星宇)

录取院校:剑桥大学(数学)

图书在版编目(CIP)数据

剑桥 A-level 纯数 1 课程精解:冲刺 G5、牛剑必备宝典:英文/雅力教育集团编著;周则鸣本册主编. —上海:复旦大学出版社,2021.3
(剑桥 A-level 国际课程备考丛书)
ISBN 978-7-309-15460-3

Ⅰ.①剑⋯　Ⅱ.①雅⋯ ②周⋯　Ⅲ.①数学-高等学校-入学考试-英国-自学参考资料　Ⅳ.①O1

中国版本图书馆 CIP 数据核字(2020)第 270131 号

剑桥 A-level 纯数 1 课程精解:冲刺 G5、牛剑必备宝典
雅力教育集团　编著
周则鸣　本册主编
责任编辑/梁　玲

复旦大学出版社有限公司出版发行
上海市国权路 579 号　邮编:200433
网址:fupnet@fudanpress.com　http://www.fudanpress.com
门市零售:86-21-65102580　团体订购:86-21-65104505
外埠邮购:86-21-65642846　出版部电话:86-21-65642845
上海四维数字图文有限公司

开本 850×1168　1/16　印张 12.75　字数 359 千
2021 年 3 月第 1 版第 1 次印刷

ISBN 978-7-309-15460-3/O·696
定价:39.00 元

如有印装质量问题,请向复旦大学出版社有限公司出版部调换。
版权所有　侵权必究